Aniruddha Datta

Adaptive Internal Model Control

With 25 Figures

Aniruddha Datta, PhD
Department of Electrical Engineering, Texas A & M University,
College Station, Texas 77843-3128, USA

ISBN 3-540-76252-3 Springer-Verlag Berlin Heidelberg New York

British Library Cataloguing in Publication Data
Datta, Aniruddha
Adaptive internal model control. - (Advances in industrial control)
1.Adaptive control systems
I.Title
629.8'36
ISBN 3540762523

Library of Congress Cataloging-in-Publication Data
Datta, Aniruddha, 1963-
Adaptive internal model control / Aniruddha Datta.
p. cm. -- (Advances in industrial control)
Includes bibliographical references.
ISBN 3-540-76252-3 (casebound : alk. paper)
1. Adaptive control systems. 2. Process control. I. Title.
II. Series.
TJ217.D36 1998
629.8'36--dc21 98-2771

Typesetting: Camera ready by author
Printed and bound at the Athenæum Press Ltd., Gateshead, Tyne & Wear
69/3830-543210 Printed on acid-free paper

Advances in Industrial Control

Springer
London
Berlin
Heidelberg
New York
Barcelona
Budapest
Hong Kong
Milan
Paris
Santa Clara
Singapore
Tokyo

Other titles published in this Series:

Modelling and Simulation of Power Generation Plants
A.W. Ordys, A.W. Pike, M.A. Johnson, R.M. Katebi and M.J. Grimble

Model Predictive Control in the Process Industry
E.F. Camacho and C. Bordons

H_∞ Aerospace Control Design: A VSTOL Flight Application
R.A. Hyde

Neural Network Engineering in Dynamic Control Systems
Edited by Kenneth Hunt, George Irwin and Kevin Warwick

Neuro-Control and its Applications
Sigeru Omatu, Marzuki Khalid and Rubiyah Yusof

Energy Efficient Train Control
P.G. Howlett and P.J. Pudney

Hierarchical Power Systems Control: Its Value in a Changing Industry
Marija D. Ilic and Shell Liu

System Identification and Robust Control
Steen Tøffner-Clausen

Genetic Algorithms for Control and Signal Processing
K.F. Man, K.S. Tang, S. Kwong and W.A. Halang

Advanced Control of Solar Plants
E.F. Camacho, M. Berenguel and F.R. Rubio

Control of Modern Integrated Power Systems
E. Mariani and S.S. Murthy

Advanced Load Dispatch for Power Systems: Principles, Practices and Economies
E. Mariani and S.S. Murthy

Supervision and Control for Industrial Processes
Björn Sohlberg

Modelling and Simulation of Human Behaviour in System Control
Pietro Carlo Cacciabue

Modelling and Identification in Robotics
Krzysztof Kozlowski

Spacecraft Navigation and Guidance
Maxwell Noton

Robust Estimation and Failure Detection
Rami Mangoubi

Advances in Industrial Control

Professor Dr -Ing M. Thoma
Institut für Regelungstechnik
Technische Universität
Appelstrasse 11
D-30167 Hannover
Germany

Professor H. Kimura
Department of Mathematical Engineering and Information Physics
Faculty of Engineering
The University of Tokyo
7-3-1 Hongo
Bunkyo Ku
Tokyo 113
Japan

Professor A.J. Laub
College of Engineering - Dean's Office
University of California
One Shields Avenue
Davis
California 95616-5294
United States of America

Professor J.B. Moore
Department of Systems Engineering
The Australian National University
Research School of Physical Sciences
GPO Box 4
Canberra
ACT 2601
Australia

Dr M.K. Masten
Texas Instruments
2309 Northcrest
Plano
TX 75075
United States of America

Professor Ton Backx
AspenTech Europe B.V.
De Waal 32
NL-5684 PH Best
The Netherlands

SERIES EDITORS' FOREWORD

The series *Advances in Industrial Control* aims to report and encourage technology transfer in control engineering. The rapid development of control technology impacts all areas of the control discipline. New theory, new controllers, actuators, sensors, new industrial processes, computer methods, new applications, new philosophies..., new challenges. Much of this development work resides in industrial reports, feasibility study papers and the reports of advanced collaborative projects. The series offers an opportunity for researchers to present an extended exposition of such new work in all aspects of industrial control for wider and rapid dissemination.

Adaptive control is one of those appealing simple ideas which has generated a wide and highly developed set of research topics. Actual implementation in industrial applications is, however, a little more problematic. It is almost as if adaptive control is a little too clever to be trusted on a routine applications basis. To overcome such industrial credibility problems a set of simple and transparent applications properties are needed. Further, industrial practitioners need to be supported by accessible, direct tutorial presentations of the component adaptive control technology components such as parameter identification and robust control.

Aniruddha Datta's monograph on Adaptive Internal Model Control makes a readable contribution to the literature on adaptive control and to the issues cited above. It seeks to present tutorial material and also develop some of the fundamental or guaranteed adaptive controller properties. The careful and consistent build-up of mathematical concepts, IMC controller structure and properties and then adding in the parameter estimation background culminating in a robust adaptive IMC scheme aids the industrial practitioner to gain a clear insight into the issues of adaptive control design and construction. The academic research community will also enjoy this full and rounded presentation of all the aspects pertinent to adaptive IMC control and may find this a useful support text for adaptive control courses.

M.J. Grimble and M.A. Johnson
Industrial Control Centre
Glasgow, Scotland, UK

To Anindita, Aparna and Anisha

PREFACE

Control systems based on the internal model control (IMC) structure are becoming increasingly popular in chemical process control industries. The IMC structure, where the controller implementation includes an explicit model of the plant, has been shown to be very effective for the control of the stable plants, typically encountered in process control. However, the implementation of a control system based on the IMC structure requires the availability of a reasonably accurate model of the plant, to be used as part of the controller. Therefore, when the plant is not accurately known, additional techniques are required to extract a model of the plant on which an IMC design can be based. One such technique is adaptive parameter estimation which can be used when the structure of the plant is reasonably well known and the predominant uncertainty is in the numerical values of the plant parameters. A controller which incorporates parameter adaptation into a control scheme based on the IMC structure is called an adaptive internal model control (AIMC) scheme.

This monograph provides a complete tutorial development of a systematic theory for the design and analysis of robust adaptive internal model control schemes. It is motivated from the fact that despite the reported industrial successes of adaptive internal model control schemes, there currently does not exist a methodology for their design with assured guarantees of stability and robustness. The ubiquitous Certainty Equivalence principle of adaptive control is invoked to combine robust adaptive laws with robust internal model controllers to obtain adaptive internal model control schemes which *can be proven* to be robustly stable. In other words, the results here provide a theoretical basis for analytically justifying some of the reported industrial successes of existing adaptive internal model control schemes. They also enable the reader to synthesize adaptive versions of his or her own favourite robust internal model control scheme by appropriately combining the latter with a robust adaptive law. This monograph will be of value to practicing engineers, researchers and graduate students interested in adaptive internal model control and its applications. It should also be of interest to adaptive control theorists who would like to see specific application areas of adaptive control theory.

It is a pleasure to acknowledge those who have helped make this monograph possible. Special appreciation is extended to Petros A. Ioannou of the

University of Southern California and Shankar P. Bhattacharyya of Texas A & M University. Petros, who was my doctoral dissertation advisor, introduced me to the topic of adaptive control in 1987 and has been instrumental in shaping my perspective on robust adaptive systems. Indeed, many of the analytical results used in this monograph for designing and analyzing robust adaptive IMC schemes originated from his group at USC, of which I was fortunate to have been a part. Shankar, on the other hand, played a pivotal role in developing my interest in the equally fascinating area of Parametric Robust Control. Appreciation is also extended to my colleagues and peers in the control community for stimulating discussions from time to time: Jo Howze, Ioannis Kanellakopoulos, Petar Kokotovic, Miroslav Krstic, Kumpati Narendra, Marios Polycarpou, Jing Sun, Gang Tao, Kostas Tsakalis, Erik Ydstie and Farid Ahmed-Zaid. I wish to thank Willy Wojsznis of Fisher-Rosemount Systems, Inc, S. Joe Qin of the University of Texas at Austin, Robert Soper, Stefani Butler, P. K. Mozumder and Gabe Barna all of Texas Instruments, Inc. for impressing upon me the industrial need for carrying out the research reported here. Financial support from the National Science Foundation and the Texas Higher Education Coordinating Board is gratefully acknowledged. Acknowledgement is also given to my current and former students: Ming-Tzu Ho, Hyo-Sik Kim, Lifford Mclauchlan, James Ochoa, Guillermo Silva and Lei Xing who either collaborated on this research, or helped with the word processing, figures, simulations, etc. besides reading through preliminary drafts of this monograph. I would finally like to thank the Series Editors M. J. Grimble and M. A. Johnson for their careful reading of the manuscript, Springer Engineering Editor Nicholas Pinefield and his Assistant Anne Neill for their support throughout the publication process, and the staff at Springer-Verlag for their expert assistance in all matters related to the camera-ready copy of the manuscript.

It is impossible to completely express my appreciation and gratitude to my family. They have provided me with all the support that I could have asked for during the time this work was being completed. I offer my thanks for their love, their encouragement, and their sacrifices, without which I could not have produced this monograph.

Aniruddha Datta

College Station, Texas June 1998

ACKNOWLEDGEMENTS

The ideas on adaptive internal model control presented in this monograph were developed over a period of several years and have previously appeared or are scheduled to appear in a number of different publications with varying degrees of completeness. The purpose of the current work is to provide a complete and unified presentation of the subject by bringing these ideas together in a single publication. To achieve this goal, it has been necessary at times to reuse some material that we reported in earlier papers. Although in most instances such material has been modified and rewritten for the monograph, permission from the following publisher is acknowledged.

We acknowledge the permission of Elsevier Science Ltd., The Boulevard, Langford Lane, Kidlington 0X5 1GB, UK to reproduce portions of the following papers.

- Datta A. and Ochoa J., "Adaptive Internal Model Control: Design and Stability Analysis," *Automatica*, Vol. 32, No. 2, 261-266, February 1996.
- Datta A. and Ochoa J., "Adaptive Internal Model Control: H_2 Optimization for Stable Plants," *Automatica*, Vol. 34, No. 1, 75-82, January 1998.

TABLE OF CONTENTS

LIST OF FIGURES

CHAPTER 1
INTRODUCTION

The control of chemical processes is an important applications area for the field of automatic control. The design of advanced control systems for chemical process control is quite a challenging task since it requires the satisfaction of several physical constraints on the values of the controlled variables. Conventional control theory is not capable of incorporating such constraints directly into the control system design. Instead, constraints are handled in a mostly adhoc fashion, once the design has been carried out. Another important characteristic of conventional automatic control theory is that a considerable amount of effort is expended in first stabilizing an open loop unstable plant. However, for most process control applications, the plant is already open loop stable to start with thereby making the initial stabilization step unnecessary. What is therefore desirable for process control applications is a control scheme which directly handles constraints and also does not destabilize a plant which is stable to start with. Model predictive control schemes [13] apparently possess both these characteristics and this is what accounts for their immense popularity in process control applications.

1.1 Model Predictive Control

Model predictive control refers to the family of control schemes where the computation of the control input at a given instant of time is carried out by using a model of the plant to predict future values of the output over a finite time horizon, and then choosing the control input to minimize a cost function involving the predicted output. The cost function is usually a quadratic one which penalizes (i) the deviations of the predicted output from the desired values and (ii) the control effort required. The calculated control input is then implemented over a time horizon which may be less than or equal to the time horizon over which the cost function was minimized. Subsequent values of the control input are obtained by repeating this procedure, starting from the instant of time when a fresh control input needs to be calculated. Constraints on process variables such as hard limits on their magnitude are readily incorporated into the optimization process, thereby accounting for the much publicized constraint handling capability of model predictive control schemes.

Model predictive control schemes such as Dynamic Matrix Control (DMC) [3] and Model Algorithmic Control (MAC) [36] became immensely popular in the petro-chemical industries in the late seventies and early eighties. In DMC, the prediction of the plant output is carried out using a truncated *step response model* while in MAC, the truncated *impulse response model* is used instead. There are a few other minor differences such as the horizon over which the control input is calculated and the actual horizon over which these calculated values are implemented, etc. A detailed treatment of these model predictive schemes and their properties is beyond the scope of this monograph; accordingly, we refer the interested reader to [13] which is a survey paper containing a fairly exhaustive list of references on the subject.

Despite their heuristic basis, model predictive schemes such as DMC and MAC have been widely and successfully used in process control applications. Their inherent constraint handling capability and the fact that process control plants are typically *open loop stable* are probably the two most important factors that have contributed to their success and widespread acceptance. However, a complete and general analysis of their properties such as stability, robustness, etc. is not possible with the currently available tools since the control laws that result are time-varying and cannot be expressed in closed form.

One of the techniques proposed as an adhoc fix to this problem is to first consider the model predictive control problem without any constraints. When the constraints are removed, the optimization problem in model predictive control becomes a standard least squares problem which can be solved quite easily [13]. Indeed, if instead of a truncated step response or impulse response model, one were to use a state space model, the resulting control law would be very similar to what is obtained in standard linear quadratic (LQ) optimal control [24]. The stability and robustness properties of unconstrained model predictive controllers can therefore be established using standard approaches. Once this is done, then instead of penalizing the deviations between the unconstrained and desired outputs in the cost function, the constrained case is handled by using the difference between the *unconstrained* and *constrained predicted outputs*. The rationale behind this adhoc fix is that if the output of the constrained plant is made to stay close to the unconstrained output, then the behaviour of the constrained closed loop system will be "close" to its *analytically established* unconstrained behaviour. A more detailed discussion on this subject can be found in [13]. Even with the limited discussion here, it is now clear that although constraint handling capability is an important feature of model predictive control, nevertheless the study of unconstrained model predictive control schemes is equally important. It is precisely in this context that the class of *internal model control* (IMC) schemes was born.

1.2 Internal Model Control

As already discussed, model predictive control schemes inherently require a model of the plant to be used for output prediction and calculation of the control input. Furthermore, when constraints on the process variables are disregarded, a linear time-invariant controller can be obtained. In an effort to combine the advantages of the various unconstrained model predictive control schemes and to avoid their disadvantages, Garcia and Morari [12] showed that most unconstrained model predictive controllers could be implemented as shown in Fig. 1.1[1] Here u and y represent the plant input and output

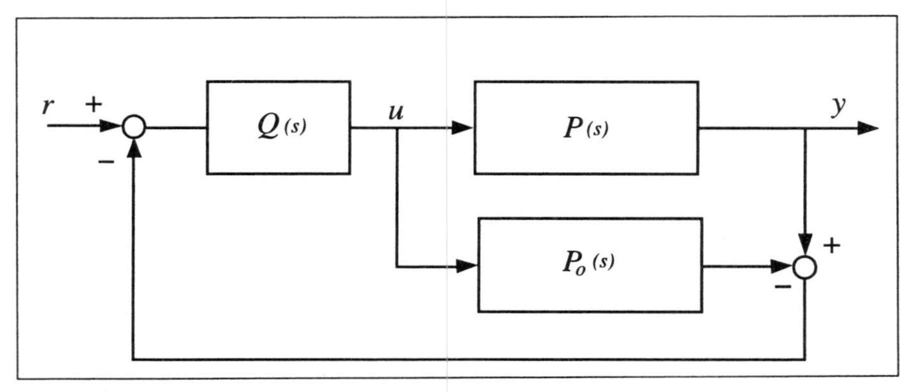

Fig. 1.1. The IMC Configuration

signals respectively; $P(s)$ represents the transfer function of the stable plant to be controlled; $P_0(s)$ is the transfer function of a plant *model* to be used for calculating the control input u; r represents the external command signal and $Q(s)$ is a stable transfer function[2] . The controller configuration shown in Fig. 1.1 is referred to as the *internal model control* (IMC) structure [30]. It can be shown that if $P(s)$ is stable and $P_0(s) = P(s)$, then the configuration in Fig. 1.1 is stable if and only if the transfer function $Q(s)$, usually referred

[1] In this monograph, our entire treatment is confined to the single-input single-output case. Although the multi-input multi-output case has been treated in [30], we prefer to stick to the single-input single-output case since adaptive results are currently available only for the latter.

[2] Implicit in the configuration of Fig. 1.1 is the notation $y = H(s)[u]$ which denotes that the time domain signal $y(t)$ is the response of the stable filter, whose transfer function is $H(s)$, to the time domain signal $u(t)$, i.e. $y(t) = h(t) * u(t)$ where '$*$' denotes time domain convolution and $h(t)$ is the impulse response of $H(s)$. Although, strictly speaking this inter-mingling of time domain signals and Laplace domain transfer functions is an abuse of notation, it has become so standard in the adaptive control literature that we will have to make repeated use of it throughout this monograph. Let us, therefore, once and for all introduce this notation and always keep in mind what it actually means.

to as the IMC parameter, is stable. In other words, if we start with an open loop stable plant and use an *exact model in the IMC design*, then the closed loop system will continue to be stable as long as the IMC parameter is chosen to be stable. This certainly is a very attractive feature from the point of view of process control applications. A detailed discussion of the IMC structure, its properties and its relationship to the classical feedback control structure is given in Chapter 3. The interested reader may also wish to consult [30] for more information on the topic.

1.3 IMC Design in the Presence of Uncertainty

The strong stability retaining property of the IMC configuration of Fig. 1.1 is conditioned on the assumption that the plant model $P_0(s)$ is an exact replica of the plant $P(s)$. However, in the real world, no plant can be modelled perfectly and so in reality $P_0(s)$ can never be equal to $P(s)$. Thus it is important to study how the mismatch between $P(s)$ and $P_0(s)$ affects the stability of the closed loop system. It would also be of interest to see if the design of the IMC parameter itself could be carried out to accomodate such a mismatch. Of course, the type of design used would very much depend on the nature of the mismatch. Two widely used design strategies for accomodating plant model mismatch are *robust control* and *adaptive control.* Although both of these strategies are generally speaking applicable to any control scheme, our discussion here will be confined to the specific case of the IMC structure.

1.3.1 Robust Control

In robust control, one seeks to design a single time-invariant controller that can assure that the closed loop system retains its properties in the presence of modelling errors such as plant model mismatch. Since the exact modelling error is, by definition, unknown, it is customary to consider classes of modelling errors that are likely to be encountered in practice. Some commonly used characterizations of modelling errors are the following.

- Multiplicative Perturbations: Here the actual plant $P(s)$ is given by $P(s) = P_0(s)[1+\Delta_m(s)]$ where $P_0(s)$ is the modelled part of the plant, and $\Delta_m(s)$ is a stable transfer function[3] such that $P_0(s)\Delta_m(s)$ is strictly proper.
- Additive Perturbations: In this case, the actual plant $P(s)$ is given by $P(s) = P_0(s) + \Delta_a(s)$ where $\Delta_a(s)$ is a stable transfer function which is strictly proper.

[3] Some of the uncertainty definitions given here are a little bit more restrictive than those given in other texts, e.g. [43]. This is because the additional restrictions imposed here will later on facilitate adaptive designs.

- Stable Factor Perturbations: In this case, the actual plant is obtained by perturbing the stable factors in a coprime factorization representation of the modelled part of the plant. A detailed description can be found in [43].
- Parametric Uncertainty: In this case, the actual plant $P(s)$ has the same structure as the model $P_0(s)$. However, the coefficients of the numerator and denominator polynomials of $P(s)$ are uncertain.
- Disturbances and Noise: These are undesirable external signals that enter at the input and output of the plant and also high frequency noise that typically accompanies sensory measurements. These signals are assumed to be bounded so that their effect can be countered using bounded controls.
- Mixed Type Uncertainty: In this case, the actual plant $P(s)$ has both parametric as well as non-parametric uncertainties; for instance, we could have $P(s) = P_{01}(s)[1 + \Delta_m(s)]$ where $P_{01}(s)$ has the same structure as $P_0(s)$ but different numerator and denominator polynomial coefficients and $\Delta_m(s)$ is a multiplicative uncertainty.

We next introduce the definitions of the *sensitivity* and *complementary sensitivity* functions which play an important role in many robust control designs. Consider the IMC structure shown in Fig. 1.1 with $P(s) = P_0(s)$. Then it is clear that the output y is related to r by

$$Y(s) = P_0(s)Q(s)R(s) \tag{1.1}$$

Also the tracking error $e(t) \stackrel{\Delta}{=} r(t) - y(t)$ satisfies

$$E(s) = [1 - P_0(s)Q(s)]R(s) \tag{1.2}$$

The transfer function $S(s) = 1 - P_0(s)Q(s)$ which relates the tracking error to the command input is called the *Sensitivity Function*. The transfer function $T(s) = P_0(s)Q(s)$ is called the *Complementary Sensitivity* function since it complements the sensitivity function in the sense that

$$S(s) + T(s) = 1 \tag{1.3}$$

As already mentioned, both S and T play an important role in robust control design. For instance, if good tracking is desired then (1.2) dictates that the sensitivity function $S(s)$ be made "small." On the other hand, if we have a multiplicative uncertainty and we want to design the IMC parameter $Q(s)$ so that the closed loop system can tolerate a larger multiplicative perturbation, then as will be shown in Chapter 2 (Example 2.3.1), $T(s)$ must be made "small." Since S and T are constrained by (1.3), the two objectives above are somewhat conflicting. However, S small is typically required in the *low frequency range* since only low frequency signals are usually tracked. At the same time, T small is usually required in the *high frequency range* where the plant model mismatch usually becomes large. Thus a compromise between these two and possibly other conflicting objectives can be achieved by choosing a stable $Q(s)$ to minimize a frequency weighted combination of $T(s)$ and

$S(s)$. H_2, H_∞ [10] and L_1 [4] optimal control have emerged as design techniques for handling such design tradeoffs. Although H_2, H_∞ and L_1 optimal control designs can be carried out for general plants which may be unstable, for the special case of open loop stable plants, the use of the IMC structure is particularly appropriate. This is because with an open-loop stable plant and the IMC structure, closed loop stability is assured as long as the IMC parameter $Q(s)$ is stable, thereby simplifying the search for the optimal $Q(s)$.

In the above mentioned optimal control design techniques, the plant model mismatch is of the *non-parametric* type and is usually modelled as norm bounded perturbations in some algebra, e.g. multiplicative perturbations. When the uncertainty is of the parametric or of the mixed type, powerful and elegant techniques are available for *analyzing* the behaviour of the closed loop system. A comprehensive discussion of these techniques which collectively constitute the area of Parametric Robust Control can be found in [1]. Unfortunately, the actual *controller design* or *synthesis* in the presence of parametric or mixed uncertainty remains a largely unsolved problem.

1.3.2 Adaptive Control

Adaptive Control is appropriate when the plant uncertainty is predominantly of the parametric type and is too large to be handled using a single fixed controller. In such a case, one can estimate on-line the numerator and denominator polynomials of the transfer function of the modelled part of the plant and design the IMC controller using these estimates. Since the parameter estimates will be changing with time, the controller will no longer be time invariant. Furthermore, since the parameter estimator is typically a nonlinear element, the closed loop system even starting with a linear plant will be nonlinear and time varying, thereby requiring some fairly intricate analysis.

The idea of combining parameter estimation with an IMC structure to obtain what are called *adaptive internal model control* (AIMC) schemes is not new. Indeed, such an idea has been used in several applications e.g. [40, 39]. What is by and large lacking in the literature, however, is the presence of any kind of theoretical guarantees for the stability, convergence, etc. of such schemes beyond what is observed in empirical simulations. The main reason for this state of affairs appears to be the fact that most adaptive control theorists are perhaps not familiar with the IMC structure, while most process control engineers lack familiarity with the subtle intricacies of adaptive control theory. Indeed, to the author's knowledge, the only available theoretical result on this problem [45] is due to an adaptive control theorist who also happens to be a process control engineer. The principal objective of the current monograph, therefore, is to narrow down this gap between theory and applications of adaptive control by presenting a systematic approach for the design and analysis of AIMC schemes. Although some papers on this topic have already been published [6, 7, 8], the in-depth tutorial coverage presented here will make the material accessible to a much wider audience including industrial practitioners who may not be familiar with the technicalities associated with adaptive system theory.

1.4 Organization of the Monograph

The contents of the monograph are organized as follows.

In Chapter 2, we present the entire mathematical background that is necessary for understanding the rest of the monograph. A reader who is familiar with the design and analysis of classical adaptive control schemes may wish to skip this chapter. Other readers, such as industrial practitioners, will find a complete tutorial coverage of the mathematical results that form the basis of adaptive system analysis and design. It should, of course, be pointed out that some of the results here are not specific to adaptive control and are frequently encountered in other areas of system theory. Nevertheless, they have been included here to make the monograph as self contained as possible.

In Chapter 3, we introduce the class of IMC schemes and discuss its relationship to the classical feedback control structure. In particular, the IMC structure is shown to be a special case of the well known Youla-Jabr-Bongiorno-Kucera (YJBK) parametrization of all stabilizing controllers [46]. Different IMC designs leading to some familiar control schemes are presented; also, the inherent robustness of the IMC structure to plant modelling errors is analyzed.

Chapter 4 considers the problem of on-line parameter estimation for a general plant in the ideal case, i.e. in the absence of modelling errors. Simple examples are used to develop the theory underlying the design of both unnormalized and normalized gradient parameter estimators. To a person unfamiliar with parameter estimation theory, this chapter provides a good tutorial introduction. However, the interested reader may wish to consult [19] for a comprehensive account of other methods for designing parameter estimators.

In Chapter 5, we consider the problem of designing IMC schemes for a plant with unknown parameters. The parameter estimators from Chapter 4 are combined with the control laws from Chapter 3 to yield adaptive IMC schemes with provable guarantees of stability and convergence in the ideal case. The detailed stability and convergence analysis is included in this chapter. Simple examples are also used to illustrate the design procedure.

Chapter 6 begins by using simple examples to show that the on-line parameter estimators of Chapter 4, and hence the adaptive IMC schemes of Chapter 5, are extremely susceptible to the presence of modelling errors. The problem is remedied by designing "robust" parameter estimators which can behave satisfactorily in the presence of modelling errors. The complete theory underlying the design of robust estimators of the gradient type is presented.

Once again, the interested reader is referred to [19] for a comprehensive discussion of other approaches for designing robust parameter estimators.

In Chapter 7, we focus on the design and analysis of robust AIMC schemes. These schemes are obtained by combining the IMC controllers of Chapter 3 with the robust parameter estimators of Chapter 6. The complete stability and robustness analysis is presented and the design procedure is illustrated using simple examples.

Chapter 8 concludes the monograph with a summary of our contributions and a discussion of the future research directions. Additionally two appendices are included. Appendix A describes the essential results needed to understand the relationship bewteen the IMC and classical feedback structures, as presented in Chapter 3. Appendix B provides an intuitive discussion of the gradient and gradient projection optimization methods used for designing parameter estimators in Chapters 4 and 6.

CHAPTER 2

MATHEMATICAL PRELIMINARIES

In this chapter, we introduce the mathematical definitions and tools that play a crucial role in the design and analysis of adaptive systems. Most of these concepts and results are also used in other areas of system theory and a reader who is familiar with them may wish to skip the relevant sections of this chapter and focus only on the items that are very specific to adaptive systems. A practicing engineer, on the other hand, will find that this chapter offers a complete tutorial coverage of all the mathematical pre-requisites that are crucial to the understanding of the rest of this monograph.

2.2 Basic Definitions

2.2.1 Positive Definite Matrices

A square matrix A with real entries is *symmetric* if $A = A^T$. A symmetric matrix A is called *positive semidefinite* if for every $x \in R^n$, $x^T Ax \geq 0$ and *positive definite* if $x^T Ax > 0 \; \forall \; x \in R^n$ with $|x| \neq 0$. It is called *negative semidefinite* (*negative definite*) if $-A$ is positive semidefinite (positive definite).

We write $A \geq 0$ if A is positive semidefinite, and $A > 0$ if A is positive definite. We write $A \geq B$ and $A > B$ if $A - B \geq 0$ and $A - B > 0$, respectively. The following lemma provides several equivalent characterizations of the positive definiteness of a symmetric matrix.

Lemma 2.2.1. *A symmetric matrix $A \in R^{n \times n}$ is positive definite if and only if any one of the following conditions holds:*
(i) $\lambda_i(A) > 0$, $i = 1, 2, \cdots, n$ where $\lambda_i(A)$ denotes the ith eigenvalue of A, which is real because $A = A^T$;
(ii) There exists a nonsingular matrix A_1 such that $A = A_1 A_1^T$;
(iii) Every principal minor of A is positive;
(iv) $x^T Ax \geq \alpha x^T x$ for some $\alpha > 0$ and $\forall x \in R^n$.

A symmetric matrix A has n orthogonal eigenvectors and can be decomposed as

$$A = U^T \Lambda U \tag{2.1}$$

where U is an orthogonal matrix (i.e. $U^T U = I$) whose columns are made up of the eigenvectors of A, and Λ is a diagonal matrix composed of the eigenvalues of A. Using (2.1), it can be shown that if $A \geq 0$, then for any vector $x \in R^n$

$$\lambda_{min}(A) x^T x \leq x^T A x \leq \lambda_{max}(A) x^T x.$$

2.2.2 Norms and L_p Spaces

We begin by introducing for vectors the analogue of the absolute value for a scalar.

Definition 2.2.1. *The* norm $|x|$ *of a vector* x *is a real valued function with the following properties:*
(i) $|x| \geq 0$ *with* $|x| = 0$ *if and only if* $x = 0$
(ii) $|\alpha x| = |\alpha||x|$ *for any scalar* α
(iii) $|x + y| \leq |x| + |y|$ *(triangle inequality)*

Example 2.2.1. For $x = [x_1, x_2, \cdots x_n]^T \in R^n$, the following are examples of some commonly used norms:

$$\begin{aligned} |x|_1 &= \sum_{i=1}^{n} |x_i| \text{ (1-norm)} \\ |x|_2 &= \left(\sum_{i=1}^{n} |x_i|^2 \right)^{\frac{1}{2}} \text{ (Euclidean norm or 2-norm)} \\ |x|_\infty &= \max_i |x_i| \text{ (infinity norm)} \end{aligned}$$

It can be verified that each of the above norms satisfies properties (i), (ii) and (iii) of Definition 2.2.1. We next proceed to define the *induced norm* of a matrix. This definition requires the notion of the *supremum*, abbreviated as 'sup', of a real valued function. The supremum of a real valued function bounded from above is its *least upper bound*. Thus, although the supremum is an upper bound, it is unique in the sense that any number smaller than the supremum cannot be an upper bound. The notion of the supremum is useful when we are trying to obtain a *tight upper bound* on a function which does not have a maximum value. This is clearly borne out by the following example.

Example 2.2.2. Consider the function $f(t) = 1 - e^{-t}$ on the interval $t \in [0, \infty)$. Clearly, this function does not have a maximum value. However, its unattainable supremum is 1. The reader can verify that 1 is indeed the *least* upper bound since for any number $x < 1$, we can always find a t such that $f(t) > x$.

When a function does have a maximum value, the supremum and the maximum coincide. We are now ready to present the definition of the induced norm of a matrix.

Definition 2.2.2. *Let* $|.|$ *be a given vector norm. Then for any matrix* $A \in R^{m\times n}$*, the quantity* $||A||$ *defined by*

$$||A|| \triangleq \sup_{x\neq 0, x\in R^n} \frac{|Ax|}{|x|} = \sup_{|x|\leq 1} |Ax| = \sup_{|x|=1} |Ax|$$

is called the induced (matrix) norm *of* A *corresponding to the vector norm* $|.|$.

The induced matrix norm satisfies the properties (i) to (iii) of Definition 2.2.1.

Some of the properties of the induced norm that will be frequently used in this monograph are summarized below:
(i) $|Ax| \leq ||A|||x| \;\forall\; x \in R^n$
(ii) $||A+B|| \leq ||A|| + ||B||$
(iii) $||AB|| \leq ||A||||B||$
where A, B are arbitrary matrices of compatible dimensions.

We next present a few examples showing how the induced norms are calculated for some commonly used vector norms.

Example 2.2.3. Show that the induced matrix norm corresponding to the vector 1-norm is given by

$$||A||_1 = \max_j \sum_{i=1}^{n} |a_{ij}| \text{ (maximum of the column sums)}$$

Proof.

$$\begin{aligned}
\text{Now } |Ax|_1 &= \sum_{i=1}^{n}\left|\sum_{j=1}^{n} a_{ij}x_j\right| \\
&\quad \text{(since the } i\text{th component of } Ax \triangleq (Ax)_i = \textstyle\sum_{j=1}^{n} a_{ij}x_j) \\
&\leq \sum_{i=1}^{n}\sum_{j=1}^{n} |a_{ij}||x_j| \\
&\quad \text{(by repeatedly using (ii) and (iii) of Definition 2.2.1)} \\
&= \sum_{j=1}^{n}\sum_{i=1}^{n} |a_{ij}||x_j| \text{ (interchanging the order of summation)} \\
&\leq [\max_j \sum_{i=1}^{n} |a_{ij}|] \sum_{j=1}^{n} |x_j|
\end{aligned}$$

$$= [\max_j \sum_{i=1}^{n} |a_{ij}|]|x|_1$$

Thus $\max_j (\sum_{i=1}^{n} |a_{ij}|)$ is an upper bound for $\frac{|Ax|_1}{|x|_1}$ provided $x \neq 0$. Since $||A||_1$ is the *least* upper bound, we must have

$$||A||_1 \leq \max_j \sum_{i=1}^{n} |a_{ij}| \tag{2.2}$$

We will now show that the less than or equal to symbol in (2.2) is in fact an equality. Towards this end, suppose that the maximum column sum occurs for $j = k$ and consider the basis vector e_k which has a 1 in the kth position and zeros everywhere else so that $|e_k|_1 = 1$.

$$\begin{aligned} \text{Now } |Ae_k|_1 &= \left(\sum_{i=1}^{n} |a_{ik}|\right) = \left(\sum_{i=1}^{n} |a_{ik}|\right) |e_k|_1 \\ \Rightarrow ||A||_1 &\geq \sum_{i=1}^{n} |a_{ik}| = \max_j \sum_{i=1}^{n} |a_{ij}| \qquad (2.3) \end{aligned}$$

Combining (2.2) and (2.3), we obtain

$$||A||_1 = \max_j \left(\sum_{i=1}^{n} |a_{ij}|\right)$$

and this completes the proof.

Example 2.2.4. Show that the induced matrix norm corresponding to the vector 2-norm is given by

$$||A||_2 = \max_i [\lambda_i(A^T A)]^{\frac{1}{2}}$$

where λ_i denotes the ith eigenvalue. The quantities $[\lambda_i(A^T A)]^{\frac{1}{2}}$, $i = 1, 2, \cdots$, n are called the *singular values* of the matrix A.

Proof. Since $(A^T A)^T = A^T A$, it is clear that $A^T A$ is a symmetric matrix. Morover, it is easy to show that $A^T A$ is positive semidefinite. Hence, it follows that

$$\begin{aligned} x^T A^T Ax &\leq \max_i \lambda_i(A^T A) x^T x \\ \Rightarrow |Ax|_2^2 &\leq \max_i \lambda_i(A^T A)(|x|_2)^2 \\ \Rightarrow |Ax|_2 &\leq [\max_i \lambda_i(A^T A)]^{\frac{1}{2}} |x|_2 \\ \Rightarrow \frac{|Ax|_2}{|x|_2} &\leq \max_i \left\{[\lambda_i(A^T A)]^{\frac{1}{2}}\right\} \text{ provided } x \neq 0 \\ \Rightarrow ||A||_2 &\leq \max_i \left\{[\lambda_i(A^T A)]^{\frac{1}{2}}\right\} \qquad (2.4) \end{aligned}$$

To show equality in the above expression, we proceed as follows. Let x_m be the eigenvector of $A^T A$ corresponding to the largest eigenvalue and assume that $|x_m|_2 = 1$.

$$\begin{aligned}
\text{Then } A^T A x_m &= [\max_i \lambda_i(A^T A)] x_m \\
\Rightarrow x_m^T A^T A x_m &= [\max_i \lambda_i(A^T A)] x_m^T x_m \\
\Rightarrow (|A x_m|_2)^2 &= [\max_i \lambda_i(A^T A)] (|x_m|_2)^2 \\
\Rightarrow |A x_m|_2 &= [\max_i \lambda_i(A^T A)]^{\frac{1}{2}} |x_m|_2 \\
\Rightarrow ||A||_2 &\geq \max_i [\lambda_i(A^T A)]^{\frac{1}{2}} \qquad (2.5)
\end{aligned}$$

The proof is now completed by combining (2.4) and (2.5).

Example 2.2.5. Show that the induced matrix norm corresponding to the vector ∞-norm is given by

$$||A||_\infty = \max_i \sum_{j=1}^{n} |a_{ij}| \text{ (maximum of the row sums)}$$

Proof.

$$\begin{aligned}
\text{Now } |Ax|_\infty &= \max_i \left| \sum_{j=1}^{n} a_{ij} x_j \right| \\
&\leq \max_i \sum_{j=1}^{n} |a_{ij}||x_j| \\
&\leq \max_i \sum_{j=1}^{n} |a_{ij}||x|_\infty \\
&= \left[\max_i \sum_{j=1}^{n} |a_{ij}| \right] |x|_\infty \\
\Rightarrow ||A||_\infty &\leq \max_i \sum_{j=1}^{n} |a_{ij}| \qquad (2.6)
\end{aligned}$$

To show equality in the above expression, we proceed as follows. Suppose that $\max_i \sum_{j=1}^{n} |a_{ij}|$ is attained for $i = k$. Define a new vector $x = (x_1, x_2, \cdots, x_n)^T$ by

$$x_j = \begin{cases} \frac{a_{kj}}{|a_{kj}|} & \text{if } a_{kj} \neq 0 \\ 1 & \text{otherwise} \end{cases}$$

Then $|x|_\infty = 1$. Now for $i = k$, $\left|\sum_{j=1}^n a_{ij}x_j\right| = \sum_{j=1}^n |a_{kj}|$ and if $i \neq k$, $\left|\sum_{j=1}^n a_{ij}x_j\right| \leq \sum_{j=1}^n |a_{ij}| < \sum_{j=1}^n |a_{kj}|$.

$$\begin{aligned}\text{Thus } |Ax|_\infty = \sum_{j=1}^n |a_{kj}| &= \left(\sum_{j=1}^n |a_{kj}|\right) |x|_\infty \\ &= \max_i \left(\sum_{j=1}^n |a_{ij}|\right) .|x|_\infty \\ \Rightarrow ||A||_\infty &\geq \max_i \left(\sum_{j=1}^n |a_{ij}|\right) \end{aligned} \tag{2.7}$$

From (2.6) and (2.7), it follows that $||A||_\infty = \max_i \left(\sum_{j=1}^n |a_{ij}|\right)$ so that the proof is complete.

For functions of time, we define the L_p norm

$$||x||_p \triangleq \left[\int_0^\infty |x(\tau)|^p d\tau\right]^{\frac{1}{p}} \tag{2.8}$$

for $p \in [1, \infty)$ and say that $x \in L_p$ when $||x||_p$ exists (i.e. when $||x||_p$ is finite). An immediate question that comes to mind is whether the $||.||_p$ defined in (2.8) really qualifies as a vector norm in accordance with Definition 2.2.1. Strictly speaking, the answer is no since one can easily find a function $x(t)$ which is not identically zero on $[0, \infty)$ for which $||x||_p$ calculated using (2.8) is zero. This should come as no surprise since changing the value of a function at a *finite* number of points does not change the value of the integral used in (2.8). To overcome this problem, we define the zero element in L_p to be the *equivalence class* made up of all those functions for which the integral in (2.8) is zero. If the value of $x(t)$ can be altered at a set of points without affecting the value of the integral in (2.8), then that set is referred to as *a set of measure zero*. A property that holds everywhere except possibly on sets of measure zero is said to hold *almost everywhere*. For instance, if we have two functions $x(t)$ and $y(t)$ defined on $[0, \infty)$ and $x(t) = y(t)$ except at a finite number of points, then we can say that $x(t) = y(t)$ almost everywhere.

The concepts of sets of measure zero and a property holding almost everywhere enable us to define the essential supremum (ess. sup) of a time function. The essential supremum of a time function is its supremum value almost everywhere. For instance consider $x : [0, 1] \mapsto R$ where $x(t) = t$, $t \in (0, 1]$ and $x(0) = 5$. Then $\sup_{t\in[0,1]} x(t) = 5$ while $ess.sup_{t\in[0,1]} x(t) = 1$. Using the notion of the essential supremum, the L_∞ norm of a time function $x(t)$ is defined as

$$||x||_\infty \triangleq ess.sup_{t\geq 0}|x(t)|$$

and we say that $x \in L_\infty$ when $||x||_\infty$ exists. In this monograph, we will write 'sup' instead of 'ess. sup' with the understanding that 'sup' denotes the 'ess. sup' whenever the two are different.

In the above L_p, L_∞ norm definitions, $x(t)$ can be a scalar or vector valued function of time. If $x(t)$ is a scalar valued function, then $|.|$ denotes the absolute value. If, on the other hand, $x(t)$ is a vector valued fuction taking values in R^n, then $|.|$ denotes any vector norm on R^n.

In control system analysis, we frequently come across signals or time functions which are not apriori known to belong to L_p. Indeed, establishing that those signals are in L_p may be one of the objectives for undertaking that analysis. To handle such functions, we define the L_{pe} norm by

$$||x_t||_p \triangleq \left[\int_0^t |x(\tau)|^p d\tau\right]^{\frac{1}{p}} \quad \text{for } p \in [1, \infty)$$

and say that $x \in L_{pe}$ when $||x_t||_p$ exists for every finite t. Similarly, the $L_{\infty e}$ norm is defined as

$$||x||_\infty \triangleq \sup_{0\leq\tau\leq t} |x(\tau)|$$

The set of all functions that belong to L_p (respectively L_{pe}) form a *vector space* called L_p space (respectively L_{pe} space).

Given a time function $f : [0, \infty) \mapsto R^n$, we can define the truncation of f to the interval $[0, t]$ by

$$f_t(\tau) \triangleq \begin{cases} f(\tau), & 0 \leq \tau \leq t \\ 0, & \tau > t \end{cases}$$

Then $f \in L_{pe} \Leftrightarrow f_t \in L_p$ for any finite t. The L_{pe} space is also called the *extended L_p space* and is defined as the set of all functions f such that $f_t \in L_p \; \forall \, t > 0$.

The next example shows that the notion of an extended space is of importance in studying the behaviour of nonlinear systems.

Example 2.2.6. Consider the nonlinear differential equation

$$\dot{x} = x^2, \; x(0) = 1 \tag{2.9}$$

The solution to this differential equation is $x(t) = \frac{1}{1-t}, \; t \in [0, 1)$ which *escapes to infinity in finite time*. Equation (2.9) is said to have *finite escape time* and $x \notin L_{pe} \; \forall \, p \in [1, \infty]$

We next present two lemmas that play an important role in the study of L_p spaces.

Lemma 2.2.2. *(Holder's Inequality) If $p, q \in [1, \infty]$ and $\frac{1}{p} + \frac{1}{q} = 1$, then $f \in L_p, g \in L_q$ imply that $fg \in L_1$ and*

$$||fg||_1 \leq ||f||_p ||g||_q$$

When $p = q = 2$, the Holder's Inequality becomes the well known Schwartz Inequality *i.e.*

$$||fg||_1 \leq ||f||_2 ||g||_2$$

Lemma 2.2.3. *(Minkowski's Inequality) For $p \in [1, \infty]$, $f, g \in L_p$ implies that $f + g \in L_p$ and*

$$||f + g||_p \leq ||f||_p + ||g||_p$$

Remark 2.2.1. The two lemmas above also hold for the truncated functions f_t, g_t of f, g respectively provided $f, g \in L_{pe}$, e.g.

$$\begin{aligned} f, g \in L_{pe} \quad &\Rightarrow \quad f_t, g_t \in L_p \\ \Rightarrow ||(fg)_t||_1 \quad &\leq \quad ||f_t||_2 ||g_t||_2 \\ \text{i.e.} \int_0^t |f(\tau)g(\tau)| d\tau \quad &\leq \quad \left[\int_0^t |f(\tau)|^2 d\tau\right]^{\frac{1}{2}} \left[\int_0^t |g(\tau)|^2 d\tau\right]^{\frac{1}{2}} \end{aligned}$$

which holds for any finite $t \geq 0$.

Example 2.2.7. Consider the time function $f(t) = \frac{1}{1+t}$, $t \geq 0$. Then $f \in L_\infty$ since $\sup_{t \geq 0} \frac{1}{1+t} = 1 < \infty$. Since $||f||_2 = \left(\int_0^\infty \frac{1}{(1+t)^2} dt\right)^{\frac{1}{2}} = 1 < \infty$, it follows that f also belongs to L_2. However, $f \notin L_1$ since $||f||_1 = \int_0^\infty \frac{1}{1+t} dt = \lim_{t \to \infty} ln(1+t) = \infty$. Nevertheless, $f \in L_{1e}$ since for any finite t, $\int_0^\infty |f_t(\tau)| d\tau = \int_0^t \frac{1}{1+\tau} d\tau = ln(1+t) < \infty$.

2.2.3 Some Properties of Functions

In the study of adaptive systems, we are quite often interested in establishing the asymptotic convergence of certain signals to zero. The establishment of such a property usually requires the use of fairly sophisticated arguments involving properties of functions of time. In this section, our objective is to discuss some of these relevant properties and to also demonstrate via simple examples that sometimes properties which may seem intuitively obvious may not even be true. This reinforces the importance of rigorous analysis in the study of adaptive systems.

Definition 2.2.3. *(Continuity) A function $f : [0, \infty) \mapsto R$ is* continuous *on $[0, \infty)$ if for any given $\epsilon > 0$, and any $t_0 \in [0, \infty)$, there exists a $\delta(\epsilon, t_0) > 0$ such that $|f(t) - f(t_0)| < \epsilon$ whenever $|t - t_0| < \delta(\epsilon, t_0)$.*

Definition 2.2.4. *(Uniform Continuity) If $\delta(\epsilon, t_0)$ in the continuity definition above does not depend on t_0, i.e. δ is a function of only ϵ, then f is said to be* uniformly continuous *on* $[0, \infty)$.

Example 2.2.8. The function $f(t) = t^2$ is continuous but not uniformly continuous on $[0, \infty)$. This can be easily checked as follows. Let $t_0 \in [0, \infty)$ be arbitrary and let $\epsilon > 0$ be given. Define $\delta(t_0, \epsilon) = \min[\sqrt{\frac{\epsilon}{2}}, \frac{\epsilon}{4t_0}]$ and suppose $t \in [0, \infty)$ such that $|t - t_0| < \delta(t_0, \epsilon)$.

$$\begin{aligned} \text{Then } |t^2 - t_0^2| &= |t + t_0||t - t_0| \\ &\leq (2t_0 + \delta)\delta \text{ (since } |t| = |t - t_0 + t_0| \leq \delta + t_0) \\ &= 2t_0\delta + \delta^2 \\ &< 2t_0 . \frac{\epsilon}{4t_0} + \frac{\epsilon}{2} = \epsilon \end{aligned}$$

Example 2.2.9. The function $f(t) = \frac{1}{1+t}$ is uniformly continuous on $[0, \infty]$. To see this, let $\epsilon > 0$ be given, and choose $\delta := \epsilon$. Now suppose $t_0, t \in [0, \infty)$ such that $|t - t_0| < \delta$. Then

$$\begin{aligned} |f(t) - f(t_0)| &= \left| \frac{1}{1+t} - \frac{1}{1+t_0} \right| \\ &= \frac{|t - t_0|}{(1+t)(1+t_0)} \\ &\leq |t - t_0| \text{ (since } 1 + t \geq 1, \ 1 + t_0 \geq 1) \\ &< \delta = \epsilon \end{aligned}$$

Lemma 2.2.4. *If $f(t)$ is differentiable on $[0, \infty)$ and $\dot{f} \in L_\infty$, then f is uniformly continuous.*

Proof. Let $\epsilon > 0$ be given and choose $\delta := \frac{\epsilon}{M}$ where $M := \sup_{t \geq 0} |\dot{f}(t)|$. Let $t_1, t_2 \in [0, \infty)$ be such that $|t_2 - t_1| < \delta$.

$$\begin{aligned} \text{Then } |f(t_2) - f(t_1)| &= |\dot{f}(\bar{t})|.|t_2 - t_1| \text{ for some } \bar{t} \text{ lying between } t_1 \text{ and } t_2 \\ &\qquad \text{(by the Intermediate Value Theorem)} \\ &< M\delta = \epsilon \end{aligned}$$

which establishes the uniform continuity of f.

Definition 2.2.5. *(Convex Set) A subset $\mathcal{K}$ of R^n is said to be* convex *if for every $x, y \in \mathcal{K}$ and for every $\alpha \in [0, 1]$, we have $\alpha x + (1 - \alpha)y \in \mathcal{K}$.*

Definition 2.2.6. *(Convex Function) A function $f : \mathcal{K} \mapsto R$ is said to be* convex *over the convex set $\mathcal{K}$ if $\forall\, x, y \in \mathcal{K}$ and $\forall\, \alpha \in [0,1]$, we have*

$$f(\alpha x + (1-\alpha)y) \leq \alpha f(x) + (1-\alpha)f(y).$$

The function $f(x) = x^2$ is convex over R because

$$\begin{aligned} f(\alpha x + (1-\alpha)y) &= \alpha^2 x^2 + (1-\alpha)^2 y^2 + 2\alpha(1-\alpha)xy \\ &\leq \alpha^2 x^2 + (1-\alpha)^2 y^2 + \alpha(1-\alpha)(x^2+y^2) \\ &\quad \text{(since } \pm 2xy \leq x^2 + y^2) \\ &= \alpha f(x) + (1-\alpha)f(y) \end{aligned}$$

We now present two examples to show that sometimes a property which may seem intuitively obvious may not even be true.

Example 2.2.10. Let $f(t) = \sin(\sqrt{1+t})$. Then $\dot{f}(t) = \frac{\cos(\sqrt{1+t})}{2\sqrt{1+t}}$. Clearly $\dot{f}(t) \to 0$ as $t \to \infty$ but $\lim_{t\to\infty} f(t)$ does not exist. Thus this example shows that if the derivative of a scalar valued function approaches zero, we cannot necessarily conclude that the function tends to a constant.

Example 2.2.11. Let $f(t) = \frac{\sin[(1+t)^n]}{(1+t)}$. Then $\dot{f}(t) = -\frac{\sin[(1+t)^n]}{(1+t)^2} + n(1+t)^{n-2}\cos[(1+t)^n]$. Thus $f(t) \to 0$ as $t \to \infty$ but $\dot{f}(t)$ has no limit as $t \to \infty$ whenever $n \geq 2$. Thus this example shows that a function converging to a constant value does not imply that the derivative converges to zero.

Lemma 2.2.5. *The following is true for scalar valued functions*
(i) A function $f(t)$ that is bounded from below and is non-increasing has a limit as $t \to \infty$.
(ii) If $0 \leq f(t) \leq g(t)\ \forall\, t \geq 0$, then for any $p \in [1,\infty]$, $g \in L_p \Rightarrow f \in L_p$.

Proof. (i) Since $f(t)$ is bounded from below, it has a largest lower bound or *infimum*[1]. Let this infimum be denoted by m so that $m := \inf_{t\in[0,\infty]} f(t)$. Let $\epsilon > 0$ be arbitrary. Since m is the *largest* lower bound for $f(t)$, $m + \epsilon$ is no longer a lower bound for $f(t)$ so that $\exists\, t_\epsilon$ such that

$$m \leq f(t_\epsilon) < m + \epsilon \tag{2.10}$$

Furthermore, since f is non-increasing, it follows that

$$f(t) \leq f(t_\epsilon)\ \forall\, t \geq t_\epsilon \tag{2.11}$$

Combining (2.10) and (2.11), we conclude that

[1] The notion of the infimum (abbreviated as 'inf') is analogous to the notion of the supremum introduced earlier.

$$\begin{aligned} m \le f(t) &< m+\epsilon \,\forall\, t \ge t_\epsilon \\ \Rightarrow m-\epsilon < f(t) &< m+\epsilon \,\forall\, t \ge t_\epsilon \\ \text{or } |f(t)-m| &< \epsilon \,\forall\, t \ge t_\epsilon \end{aligned}$$

Thus $\forall\, \epsilon > 0$, $\exists\, t_\epsilon$ such that $t \ge t_\epsilon \Rightarrow |f(t)-m| < \epsilon$. This is equivalent to $\lim_{t\to\infty} f(t) = m$.
(ii) For $p=\infty$, the proof is trivial. So, let us consider the case $p \in [1,\infty)$. Then

$$\begin{aligned} \|f_t\|_p &= \left[\int_0^t |f(\tau)|^p d\tau\right]^{\frac{1}{p}} \\ &\le \left[\int_0^\infty |g(\tau)|^p d\tau\right]^{\frac{1}{p}} \quad (\text{since } 0 \le f(t) \le g(t)\ ) \end{aligned}$$

Since $\|f_t\|_p$ is monotonically non-decreasing and bounded from above, it follows from the result analogous to (i) that $\lim_{t\to\infty} \|f_t\|_p$ exists so that $f \in L_p$.

Lemma 2.2.6. *(Differential Inequalities) Let $f, V : [0,\infty) \mapsto R$. Then*

$$\dot{V} \le -\alpha V + f,\ \forall\, t \ge t_0 \ge 0$$

implies that

$$V(t) \le e^{-\alpha(t-t_0)} V(t_0) + \int_{t_0}^t e^{-\alpha(t-\tau)} f(\tau) d\tau,\ \forall\, t \ge t_0 \ge 0$$

for any finite constant α.

Proof. Since $\dot{V} \le -\alpha V + f$, $\forall\, t \ge t_0 \ge 0$, there exists a function $z(t) \ge 0$ for all $t \ge t_0 \ge 0$ such that

$$\dot{V} = -\alpha V + f - z$$

Writing down the solution of the above differential equation, we obtain

$$\begin{aligned} V(t) &= e^{-\alpha(t-t_0)} V(t_0) + \int_{t_0}^t e^{-\alpha(t-\tau)} [f(\tau) - z(\tau)] d\tau \\ &= e^{-\alpha(t-t_0)} V(t_0) + \int_{t_0}^t e^{-\alpha(t-\tau)} f(\tau) d\tau \\ &\quad - \int_{t_0}^t e^{-\alpha(t-\tau)} z(\tau) d\tau \\ &\le e^{-\alpha(t-t_0)} V(t_0) + \int_{t_0}^t e^{-\alpha(t-\tau)} f(\tau) d\tau \\ &\qquad (\text{since } \textstyle\int_{t_0}^t e^{-\alpha(t-\tau)} z(\tau) d\tau \ge 0) \end{aligned}$$

and this completes the proof.

Lemma 2.2.7. *(Barbalat's Lemma) If* $\lim_{t\to\infty}\int_0^t f(\tau)d\tau$ *exists and is finite, and* $f(t)$ *is a uniformly continuous function of time, then* $\lim_{t\to\infty} f(t) = 0$.

Proof. By way of contradiction, let us assume that $\lim_{t\to\infty} f(t) = 0$ does not hold, i.e. either the limit does not exist or is not equal to zero. This implies that there exists an $\epsilon_0 > 0$ such that for every $T > 0$, one can find $t_i \geq T$ such that $|f(t_i)| > \epsilon_0$. Since f is uniformly continuous, there is a positive constant $\delta(\epsilon_0)$ such that $|f(t) - f(t_i)| < \frac{\epsilon_0}{2}\ \forall\, t \in [t_i, t_i + \delta(\epsilon_0)]$. Hence, for all $t \in [t_i, t_i + \delta(\epsilon_0)]$, we have

$$\begin{aligned} |f(t)| &= |f(t) - f(t_i) + f(t_i)| \\ &\geq |f(t_i)| - |f(t) - f(t_i)| \\ &> \epsilon_0 - \frac{\epsilon_0}{2} = \frac{\epsilon_0}{2} \end{aligned}$$

which implies that

$$\left| \int_{t_i}^{t_i+\delta(\epsilon_0)} f(\tau)d\tau \right| = \int_{t_i}^{t_i+\delta(\epsilon_0)} |f(\tau)|d\tau > \frac{\epsilon_0\delta(\epsilon_0)}{2} \tag{2.12}$$

where the first equality holds since $f(t)$ does not change sign on $[t_i, t_i+\delta(\epsilon_0)]$. On the other hand, $g(t) \triangleq \int_0^t f(\tau)d\tau$ has a limit as $t \to \infty$ implies that $\exists\, T_1(\epsilon_0)$ such that $\forall\, t_2 > t_1 > T_1(\epsilon_0)$ we have

$$\begin{aligned} |g(t_1) - g(t_2)| &< \epsilon_0 \frac{\delta(\epsilon_0)}{2} \\ \text{i.e.} \left| \int_{t_1}^{t_2} f(\tau)d\tau \right| &< \epsilon_0 \frac{\delta(\epsilon_0)}{2} \end{aligned} \tag{2.13}$$

Choosing $T = T_1(\epsilon_0)$, $t_1 = t_i$, $t_2 = t_i + \delta(\epsilon_0)$, we see that (2.12) and (2.13) contradict each other and, therefore, $\lim_{t\to\infty} f(t) = 0$.

Corollary 2.2.1. *If* $f, \dot{f} \in L_\infty$ *and* $f \in L_p$ *for some* $p \in [1, \infty)$, *then* $f(t) \to 0$ *as* $t \to \infty$.

Proof. Now $\frac{d}{dt}[f^p(t)] = pf^{p-1}\dot{f} \in L_\infty$ since $f, \dot{f} \in L_\infty$. Thus $f^p(t)$ is uniformly continuous. Furthermore, since $f \in L_p$, we have

$$\left| \int_0^\infty f^p(\tau)d\tau \right| \leq \int_0^\infty |f(\tau)|^p d\tau < \infty$$

Hence, by Barbalat's Lemma, it follows that $\lim_{t\to\infty} f^p(t) = 0$ which in turn implies that $\lim_{t\to\infty} f(t) = 0$.

The following example demonstrates that the uniform continuity assumption in Lemma 2.2.7 is a crucial one.

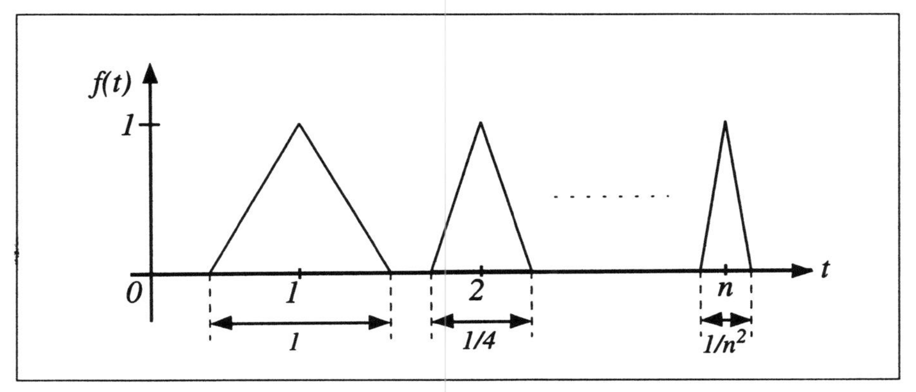

Fig. 2.1. A function which is continuous but not uniformly continuous

Example 2.2.12. Consider the following function described by a sequence of isosceles triangles of base length $\frac{1}{n^2}$ and height equal to 1 centered at n where $n = 1, 2, \cdots, \infty$ as shown in Fig. 2.1. This function is continuous but not uniformly continuous. It satisfies

$$\lim_{t\to\infty} \int_0^t f(\tau)d\tau = \frac{1}{2}\sum_{n=1}^{\infty} \frac{1}{n^2} < \infty$$

but $\lim_{t\to\infty} f(t)$ does not exist.

2.3 Input-output Stability

The systems encountered in this monograph can be described by an input/output (I/O) mapping that assigns to each input a corresponding output, or by a state variable representation. In this section, we present some basic results concerning I/O stability. Similar results for the state variable representation will be presented in Section 2.4.

2.3.1 L_p Stability

We consider a linear time invariant (LTI) causal system described by the convolution of two time functions $u, h : R^+ \mapsto R$ defined as

$$y(t) = h(t) * u(t) \triangleq \int_0^t h(t-\tau)u(\tau)d\tau = \int_0^t u(t-\tau)h(\tau)d\tau \qquad (2.14)$$

where u, y represent the input and output signals respectively. Let $H(s)$ be the Laplace transform of $h(t)$. $H(s)$ is called the transfer function of (2.14) and $h(t)$ is called its impulse response. Using the transfer function $H(s)$, the system (2.14) may also be represented in the form

$$Y(s) = H(s)U(s) \tag{2.15}$$

where $U(s), Y(s)$ represent the Laplace transforms of $u(t), y(t)$ respectively.

We say that the system represented by (2.14) or (2.15) is *L_p stable* if $u \in L_p \Rightarrow y \in L_p$ and $||y||_p \leq c||u||_p$ for some constant $c \geq 0$ and any $u \in L_p$. When $p = \infty$, L_p stability, i.e., L_∞ stability, is also referred to as bounded-input bounded-output *(BIBO) stability.*

The following results hold for the system (2.14).

Theorem 2.3.1. *If $u \in L_{pe}$ and $h \in L_1$, then*

$$||y_t||_p \leq ||h||_1 ||u_t||_p \ \forall\, t \in [0, \infty] \text{ and } \forall\, p \in [1, \infty] \tag{2.16}$$

Proof. For $p \in [1, \infty)$, we have

$$\begin{aligned} |y(t)| &\leq \int_0^t |h(t-\tau)||u(\tau)|d\tau \\ &= \int_0^t |h(t-\tau)|^{\frac{p-1}{p}} |h(t-\tau)|^{\frac{1}{p}} |u(\tau)| d\tau \\ &\leq \left(\int_0^t |h(t-\tau)|d\tau\right)^{\frac{p-1}{p}} \left(\int_0^t |h(t-\tau)||u(\tau)|^p d\tau\right)^{\frac{1}{p}} \\ &= ||h_t||_1^{\frac{p-1}{p}} \left(\int_0^t |h(t-\tau)||u(\tau)|^p d\tau\right)^{\frac{1}{p}} \end{aligned} \tag{2.17}$$

where the second inequality is obtained by applying Holder's inequality. Raising (2.17) to the power p and integrating from 0 to t, we get

$$\begin{aligned} (||y_t||_p)^p &\leq \int_0^t ||h_t||_1^{p-1} \left(\int_0^\tau |h(\tau - s)||u(s)|^p ds\right) d\tau \\ &= ||h_t||_1^{p-1} \int_0^t \left(\int_s^t |h(\tau - s)||u(s)|^p d\tau\right) ds \\ &\qquad \text{(interchanging the order of integration)} \\ &= ||h_t||_1^{p-1} \int_0^t \left(\int_0^t |h(\tau - s)||u(s)|^p d\tau\right) ds \\ &\qquad \text{(since, by causality, } h(t) = 0 \text{ for } t < 0) \\ &= ||h_t||_1^{p-1} \int_0^t |u(s)|^p \left(\int_0^t |h(\tau - s)| d\tau\right) ds \\ &\leq ||h_t||_1^{p-1} \int_0^t |u(s)|^p \left(\int_0^t |h(\tau)| d\tau\right) ds \\ &\leq ||h_t||_1^{p-1} . ||h_t||_1^1 . ||u_t||_p^p \\ &= ||h_t||_1^p . ||u_t||_p^p \\ &\leq ||h||_1^p . ||u_t||_p^p \end{aligned}$$

$$\Rightarrow ||y_t||_p \quad \leq \quad ||h||_1.||u_t||_p$$

This completes the proof for the case $p \in [1, \infty)$. The proof for the case $p = \infty$ is immediate by taking the supremum of u over $[0, t]$ in the convolution.

When $p = 2$, we have a sharper bound for $||y_t||_p$ than that of (2.16) given by the following lemma.

Lemma 2.3.1. *If $u \in L_{2e}$ and $h \in L_1$, then*

$$||y_t||_2 \leq \sup_{w \in R} |H(j\omega)|.||u_t||_2 \tag{2.18}$$

Proof. Since $h \in L_1$ and $u \in L_{2e}$, from Theorem 2.3.1, it follows that $y \in L_{2e}$. Furthermore,

$$\begin{aligned}
||y_t||_2^2 &= ||(h * u)_t||_2^2 \\
&= ||(h * u_t)_t||_2^2 \text{ (by causality)} \\
&\leq ||(h * u_t)||_2^2 \\
&= \frac{1}{2\pi}\int_{-\infty}^{\infty} [H(j\omega)U_t(j\omega)][H(j\omega)U_t(j\omega)]^* d\omega \\
&\quad \text{(using Parseval's Theorem [22])} \\
&= \frac{1}{2\pi}\int_{-\infty}^{\infty} |H(j\omega)|^2 |U_t(j\omega)|^2 d\omega \\
&\leq [\sup_{w \in R} |H(j\omega)|]^2 \left(||u_t||_2\right)^2 \text{ (using Parseval's Theorem again)}
\end{aligned}$$

This completes the proof.

Remark 2.3.1. The quantity $\sup_{\omega \in R} |H(j\omega)|$ in (2.18) is called the H_∞ *norm* of the transfer function $H(s)$ and is denoted by

$$||H(s)||_\infty \stackrel{\Delta}{=} \sup_{\omega \in R} |H(j\omega)|$$

As shown in [9], the H_∞ norm of a stable transfer function is its induced L_2 norm.

Remark 2.3.2. It can be shown [44] that (2.18) also holds when $h(.)$ is of the form

$$h(t) = \begin{cases} 0 & t < 0 \\ \sum_{i=0}^{\infty} h_i \delta(t - t_i) + h_a(t) & t \geq 0 \end{cases}$$

where $h_a \in L_1$, $\sum_{i=0}^{\infty} |h_i| < \infty$ and t_i are finite constants. The Laplace transform of $h(t)$ is now given by $H(s) = \sum_{i=0}^{\infty} h_i e^{-t_i s} + H_a(s)$ which is not a rational function of s. The biproper transfer functions that are of interest in this monograph belong to this class.

Let us now consider the case when $h(t)$ in (2.14) is the impulse response of a linear time invariant system whose transfer function $H(s)$ is a *rational* function of s. Then the following theorems hold [9].

Theorem 2.3.2. *Let $H(s)$ be a strictly proper rational function of s. Then $H(s)$ is analytic in $Re[s] \geq 0$ if and only if $h \in L_1$.*

Theorem 2.3.3. *If $h \in L_1$, then*
(i) h decays exponentially, i.e. $|h(t)| \leq \alpha_1 e^{-\alpha_0 t}$ for some $\alpha_1, \alpha_0 > 0$
(ii) $u \in L_1 \Rightarrow y \in L_1 \cap L_\infty$, $\dot{y} \in L_1$, y is continuous and $\lim_{t\to\infty} |y(t)| = 0$.
(iii) $u \in L_2 \Rightarrow y \in L_2 \cap L_\infty$, $\dot{y} \in L_2$, y is continuous and $\lim_{t\to\infty} |y(t)| = 0$.
(iv) For $p \in [1, \infty]$, $u \in L_p \Rightarrow y, \dot{y} \in L_p$ and y is continuous.

Definition 2.3.1. *(μ small in the mean square sense (m.s.s.)) Let $x : [0, \infty) \mapsto R^n$, where $x \in L_{2e}$, and consider the set*

$$\mathcal{S}(\mu) = \left\{ x : [0, \infty) \mapsto R^n \mid \int_t^{t+T} x^T(\tau) x(\tau) d\tau \leq c_0 \mu T + c_1, \ \forall\, t, T \geq 0 \right\}$$

for a given constant $\mu \geq 0$, where $c_0, c_1 \geq 0$ are some finite constants, and c_0 is independent of μ. We say that x is μ-small in the m.s.s. if $x \in \mathcal{S}(\mu)$.

Using the preceding definition, we can obtain a result similar to that of Theorem 2.3.3 (iii) in the case where $u \notin L_2$ but $u \in \mathcal{S}(\mu)$ for some constant $\mu \geq 0$.

Lemma 2.3.2. *Consider the system (2.14). If $h \in L_1$, then $u \in \mathcal{S}(\mu)$ implies that $y \in L_\infty$ for any finite $\mu \geq 0$.*

Proof. Using Theorem 2.3.3(i), we have

$$\begin{aligned} |y(t)|^2 &\leq \left(\int_0^t \alpha_1 e^{-\alpha_0 (t-\tau)} |u(\tau)| d\tau \right)^2 \ \forall\, t \geq 0 \\ &\leq \alpha_1^2 \left(\int_0^t e^{-\alpha_0 (t-\tau)} d\tau \right) \left(\int_0^t e^{-\alpha_0 (t-\tau)} |u(\tau)|^2 d\tau \right) \\ &\qquad \text{(using the Schwartz Inequality)} \\ &\leq \frac{\alpha_1^2}{\alpha_0} \int_0^t e^{-\alpha_0 (t-\tau)} |u(\tau)|^2 d\tau \end{aligned}$$

$$\begin{aligned} \text{Now } \Delta(t, 0) &\triangleq \int_0^t e^{-\alpha_0 (t-\tau)} |u(\tau)|^2 d\tau \\ &\leq e^{-\alpha_0 t} \sum_{i=0}^{n} \int_i^{i+1} e^{\alpha_0 \tau} |u(\tau)|^2 d\tau \\ &\leq e^{-\alpha_0 t} \sum_{i=0}^{n} e^{\alpha_0 (i+1)} \int_i^{i+1} |u(\tau)|^2 d\tau \end{aligned}$$

where n is an integer which satisfies $n \le t < n+1$. Since $u \in \mathcal{S}(\mu)$, it follows that

$$\begin{aligned}\Delta(t,0) &\le e^{-\alpha_0 t}(c_0\mu + c_1)\sum_{i=0}^{n} e^{\alpha_0(i+1)} \\ &\le \frac{(c_0\mu + c_1)e^{-\alpha_0 t}e^{\alpha_0(n+1)}}{1-e^{-\alpha_0}} \\ &\le \frac{(c_0\mu + c_1)e^{\alpha_0}}{1-e^{-\alpha_0}} < \infty\end{aligned}$$

Thus $y \in L_\infty$ and the proof is complete.

Definition 2.3.1 may be generalized to the case where μ is not necessarily a constant as follows.

Definition 2.3.2. *Let $x : [0,\infty) \mapsto R^n$, $w : [0,\infty) \mapsto R^+$ where $x \in L_{2e}$, $w \in L_{1e}$ and consider the set*

$$\mathcal{S}(w) = \left\{ x \middle| \int_t^{t+T} x^T(\tau)x(\tau)d\tau \le c_0 \int_t^{t+T} w(\tau)d\tau + c_1,\ \forall\, t, T \ge 0 \right\}$$

where $c_0, c_1 \ge 0$ are some finite constants. We say that x is w-small in the m.s.s. if $x \in \mathcal{S}(w)$.

Lemma 2.3.2 along with definitions 2.3.1 and 2.3.2 are repeatedly used in Chapters 6 and 7 for the analysis of the robustness properties of parameter estimators and robust adaptive IMC schemes.

2.3.2 The $L_{2\delta}$ Norm and I/O Stability

The definitions and results of the last section enable us to develop I/O stability results based on a different norm that is particularly useful in the analysis of adaptive systems. This norm is the *exponentially weighted L_2 norm* defined as

$$\|x_t\|_2^\delta \triangleq \left(\int_0^t e^{-\delta(t-\tau)} x^T(\tau)x(\tau)d\tau \right)^{\frac{1}{2}}$$

where $\delta \ge 0$ is a constant. We say that $x \in L_2^\delta$ if $\|x_t\|_2^\delta$ exists. When $\delta = 0$, we omit it from the superscript and use the notation $x \in L_{2e}$. We refer to $\|(.)_t\|_2^\delta$ as the L_2^δ norm. For any finite t, the L_2^δ norm satisfies the norm properties given in Definition 2.2.1, i.e.
(i) $\|x_t\|_2^\delta \ge 0$ with $\|x_t\|_2^\delta = 0$ if and only if $x = 0$ in the sense that it belongs to the zero equivalence class.
(ii) $\|(\alpha x)_t\|_2^\delta = |\alpha|\|x_t\|_2^\delta$ for any constant scalar α
(iii) $\|(x+y)_t\|_2^\delta \le \|x_t\|_2^\delta + \|y_t\|_2^\delta$.

Let us consider the linear time invariant system given by

$$y = H(s)[u] \tag{2.19}$$

where $H(s)$ is a rational function of s and examine L_2^δ *stability*, i.e. given $u \in L_2^\delta$, what can we say about the L_p, L_2^δ properties of the output $y(t)$ and the related bounds.

Lemma 2.3.3. *Let $H(s)$ in (2.19) be proper. If $H(s)$ is analytic in $Re[s] \geq -\frac{\delta}{2}$ for some $\delta \geq 0$ and $u \in L_{2e}$ then*
(i) $\|y_t\|_2^\delta \leq \|H(s)\|_\infty^\delta . \|u_t\|_2^\delta$
where $\|H(s)\|_\infty^\delta \stackrel{\Delta}{=} \sup_{\omega \in R} |H(j\omega - \frac{\delta}{2})|$
(ii) Furthermore, when $H(s)$ is strictly proper, we have

$$\begin{aligned} |y(t)| &\leq \|H(s)\|_2^\delta . \|u_t\|_2^\delta \\ \textit{where } \|H(s)\|_2^\delta &\stackrel{\Delta}{=} \frac{1}{\sqrt{2\pi}} \left\{ \int_{-\infty}^{\infty} |H(j\omega - \frac{\delta}{2})|^2 d\omega \right\}^{\frac{1}{2}} \end{aligned}$$

Proof. The transfer function $H(s)$ can be expressed as $H(s) = d + H_a(s)$ with[2]

$$h(t) = \begin{cases} 0, & t < 0 \\ d\delta_\Delta(t) + h_a(t), & t \geq 0 \end{cases}$$

Because d is a finite constant, $H(s)$ being analytic in $\text{Re}[s] \geq -\frac{\delta}{2}$ implies that $h_a \in L_1$ i.e. the pair $\{H(s), h(t)\}$ belongs to the class of functions considered in Remark 2.3.2. If we define

$$h_\delta(t) = \begin{cases} 0, & t < 0 \\ d\delta_\Delta(t) + e^{\frac{\delta}{2}t} h_a(t), & t \geq 0 \end{cases},$$

$y_\delta(t) \stackrel{\Delta}{=} e^{\frac{\delta}{2}t} y(t)$ and $u_\delta(t) \stackrel{\Delta}{=} e^{\frac{\delta}{2}t} u(t)$, it follows from (2.14) that

$$y_\delta(t) = \int_0^t e^{\frac{\delta}{2}(t-\tau)} h(t-\tau) e^{\frac{\delta}{2}\tau} u(\tau) d\tau = h_\delta * u_\delta$$

Now $u \in L_{2e} \Rightarrow u_\delta \in L_{2e}$. Since $H(s)$ is analytic in $\text{Re}[s] \geq -\frac{\delta}{2}$, it follows that $H(s - \frac{\delta}{2})$ is analytic in $\text{Re}[s] \geq 0$ which in turn implies that $h_a(t) e^{\frac{\delta}{2}t} \in L_1$. Hence $h_\delta(t)$ is of the form considered in Remark 2.3.2 and so, using Lemma 2.3.1 and noting that $H(s - \frac{\delta}{2})$ is the Laplace transform of $h_\delta(t)$, we obtain

$$\|(y_\delta)_t\|_2 \leq \|H(s - \frac{\delta}{2})\|_\infty \|(u_\delta)_t\|_2 \tag{2.20}$$

Since $\|y_t\|_2^\delta = e^{-\frac{\delta}{2}t} \|(y_\delta)_t\|_2$, $\|u_t\|_2^\delta = e^{-\frac{\delta}{2}t} \|(u_\delta)_t\|_2$ and $\|H(s)\|_\infty^\delta = \|H(s - \frac{\delta}{2})\|_\infty$, (i) follows directly from (2.20). We now turn to the proof of (ii).

When $H(s)$ is strictly proper, we have

[2] Here $\delta_\Delta(t)$ is used to represent the Dirac Delta function. This is done for the purpose of avoiding any possible confusion since here we do also have a scalar δ.

$$
\begin{aligned}
|y(t)| &\leq \left| \int_0^t e^{\frac{\delta}{2}(t-\tau)} h(t-\tau) e^{-\frac{\delta}{2}(t-\tau)} u(\tau) d\tau \right| \\
&\leq \left[\int_0^t e^{\delta(t-\tau)} |h(t-\tau)|^2 d\tau \right]^{\frac{1}{2}} \|u_t\|_2^\delta \\
&\qquad \text{(using the Schwartz Inequality)} \\
&\leq \left[\int_0^\infty e^{\delta\tau} |h(\tau)|^2 d\tau \right]^{\frac{1}{2}} \|u_t\|_2^\delta \\
&= \frac{1}{\sqrt{2\pi}} \left(\int_{-\infty}^{\infty} |H(j\omega - \frac{\delta}{2})|^2 d\omega \right)^{\frac{1}{2}} . \|u_t\|_2^\delta \\
&\qquad \text{(using Parseval's Theorem)} \\
&= \|H(s)\|_2^\delta . \|u_t\|_2^\delta
\end{aligned}
$$

and this completes the proof of (ii).

We refer to $\|H(s)\|_2^\delta$ and $\|H(s)\|_\infty^\delta$ defined in Lemma 2.3.3 as the δ-shifted H_2 and H_∞ norms respectively. Here it is appropriate to point out that for a strictly proper stable rational transfer function $H(s)$, the H_2 norm is defined as

$$
\|H(s)\|_2 \triangleq \frac{1}{\sqrt{2\pi}} \left\{ \int_{-\infty}^{\infty} |H(j\omega)|^2 d\omega \right\}^{\frac{1}{2}}
$$

Lemma 2.3.4. *Consider the linear time varying system given by*

$$
\dot{x} = A(t)x + B(t)u, \; x(0) = x_0 \tag{2.21}
$$

where $x \in R^n$, $u \in R^m$, and the matrices $A, B \in L_\infty$ and their elements are continuous functions of time. If the state transition matrix[3] *$\Phi(t,\tau)$ of (2.21) satisfies:*

$$
\|\Phi(t,\tau)\| \leq c_0 e^{-p_0(t-\tau)} \tag{2.22}
$$

for some $c_0, p_0 > 0$ and $u(t)$ is continuous, then for any $\delta \in [0, p_0)$, we have
(i) $|x(t)| \leq \frac{\lambda}{\sqrt{p_0}} \|u_t\|_2^\delta + \epsilon_t$
(ii) $\|x_t\|_2^\delta \leq \frac{\lambda}{\sqrt{p_0(p_0-\delta)}} \|u_t\|_2^\delta + \epsilon_t$
where ϵ_t is an exponentially decaying to zero term due to x_0 and $\lambda = cc_0$ where c is the L_∞ bound for $B(t)$.

Proof. The solution $x(t)$ of (2.21) can be expressed as

$$
x(t) = \Phi(t,0)x_0 + \int_0^t \Phi(t,\tau)B(\tau)u(\tau)d\tau.
$$

Therefore

[3] The state transition matrix $\Phi(t,\tau)$ of (2.21) is the time-varying matrix which satisfies the differential equation $\frac{\partial}{\partial t}\Phi(t,\tau) = A(t)\Phi(t,\tau)$, $\forall\, t \geq \tau$, $\Phi(\tau,\tau) = I$.

$$|x(t)| \leq \|\Phi(t,0)\|.|x_0| + \int_0^t \|\Phi(t,\tau)\|.\|B(\tau)\||u(\tau)|d\tau.$$

Using (2.22), we have

$$|x(t)| \leq \epsilon_t + \lambda \int_0^t e^{-p_0(t-\tau)}|u(\tau)|d\tau \tag{2.23}$$

where c and λ are as defined in the statement of the lemma. Applying the Schwartz inequality and using the fact that $0 \leq \delta < p_0$, we have

$$\begin{aligned} |x(t)| &\leq \epsilon_t + \lambda \left(\int_0^t e^{-p_0(t-\tau)} d\tau \right)^{\frac{1}{2}} . \left(\int_0^t e^{-\delta(t-\tau)} |u(\tau)|^2 d\tau \right)^{\frac{1}{2}} \\ &\leq \epsilon_t + \frac{\lambda}{\sqrt{p_0}} \|u_t\|_2^\delta \end{aligned}$$

which completes the proof of (i). Using the triangular property of the $\|(.)_t\|_2^\delta$-norm, it follows from (2.23) that

$$\|x_t\|_2^\delta \leq \|\epsilon_t\|_2^\delta + \lambda \|Q_t\|_2^\delta \tag{2.24}$$

$$\begin{aligned} \text{where } \|Q_t\|_2^\delta &\triangleq \left\| \left(\int_0^t e^{-p_0(t-\tau)} |u(\tau)| d\tau \right)_t \right\|_2^\delta \\ &= \left[\int_0^t e^{-\delta(t-\tau)} \left(\int_0^\tau e^{-p_0(\tau-s)} |u(s)| ds \right)^2 d\tau \right]^{\frac{1}{2}} \end{aligned}$$

Using the Schwartz inequality, we have:

$$\begin{aligned} \|Q_t\|_2^\delta &\leq \left[\int_0^t e^{-\delta(t-\tau)} \left(\int_0^\tau e^{-p_0(\tau-s)} ds \right. \right. \\ &\qquad \left. \left. . \int_0^\tau e^{-p_0(\tau-s)} |u(s)|^2 ds \right) d\tau \right]^{\frac{1}{2}} \\ &\leq \frac{1}{\sqrt{p_0}} \left[\int_0^t e^{-\delta(t-\tau)} \int_0^\tau e^{-p_0(\tau-s)} |u(s)|^2 ds d\tau \right]^{\frac{1}{2}} \end{aligned}$$

where the last inequality is obtained from the use of $\int_0^\tau e^{-p_0(\tau-s)}ds \leq \frac{1}{p_0}$. We now change the sequence of integration and obtain

$$\begin{aligned} \|Q_t\|_2^\delta &\leq \frac{1}{\sqrt{p_0}} \left[e^{-\delta t} \int_0^t e^{p_0 s} |u(s)|^2 . \left(\int_s^t e^{-(p_0-\delta)\tau} d\tau \right) ds \right]^{\frac{1}{2}} \\ &\leq \frac{1}{\sqrt{p_0}} \left[e^{-\delta t} \int_0^t e^{p_0 s} |u(s)|^2 \frac{e^{-(p_0-\delta)s}}{p_0 - \delta} ds \right]^{\frac{1}{2}} \\ &\leq \frac{1}{\sqrt{p_0(p_0-\delta)}} \|u_t\|_2^\delta \end{aligned}$$

which together with (2.24) and the fact that $\|\epsilon_t\|_2^\delta$ is also exponentially decaying to zero complete the proof.

2.3.3 The Small Gain Theorem

Many feedback systems, including adaptive control systems can be put in the form shown in Fig. 2.2 The operators H_1, H_2 act on e_1, e_2 to produce the

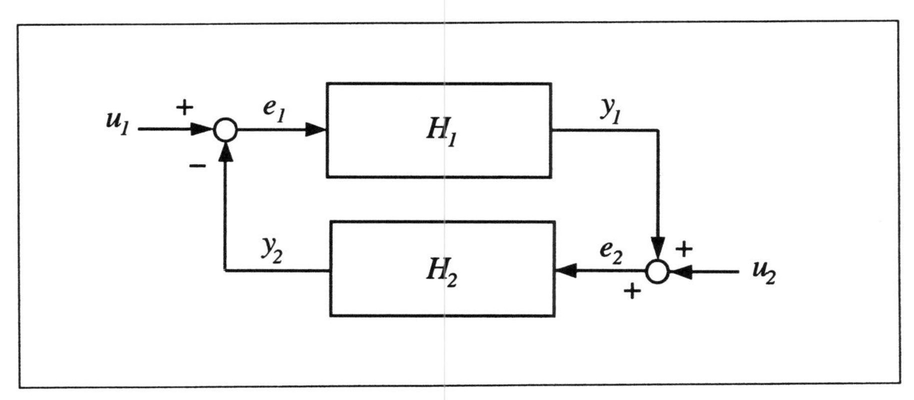

Fig. 2.2. Feedback System

outputs y_1, y_2 while u_1 and u_2 are the external inputs. Sufficient conditions for H_1, H_2 to guarantee the existence and uniqueness of the solutions e_1, y_1, e_2, y_2 for given inputs $u_1, u_2 \in L_{pe}$ are discussed in [9]. Here, we assume that H_1, H_2 are such that the existence and uniqueness of solutions is guaranteed, and proceed to determine conditions on H_1, H_2 so that if u_1, u_2 are bounded in some sense, then e_1, e_2, y_1, y_2 are also bounded in the same sense.

Let $\mathcal{L}$ be a normed linear space defined by

$$\mathcal{L} \triangleq \{f : R^+ \mapsto R^n \mid \|f\| < \infty\}$$

where $\|.\|$ corresponds to any of the L_p norms introduced earlier. Let $\mathcal{L}_e$ be the extended normed space associated with $\mathcal{L}$, i.e.

$$\mathcal{L}_e = \{f : R^+ \mapsto R^n \mid \|f_t\| < \infty, \forall\, t \in R^+\}$$

The following theorem known as the *small gain theorem* [9] gives sufficient conditions under which bounded inputs produce bounded outputs in the feedback system of Fig. 2.2.

Theorem 2.3.4. *Consider the system shown in Fig. 2.2. Suppose $H_1, H_2 : \mathcal{L}_e \mapsto \mathcal{L}_e$; $e_1, e_2 \in \mathcal{L}_e$. Furthermore, suppose that there exist constants $\gamma_1, \gamma_2 \geq 0$ and β_1, β_2 such that the operators H_1 and H_2 satisfy:*

$$\|(H_1 e_1)_t\| \leq \gamma_1 \|e_{1t}\| + \beta_1 \tag{2.25}$$
$$\|(H_2 e_2)_t\| \leq \gamma_2 \|e_{2t}\| + \beta_2 \tag{2.26}$$

$\forall\, t \in R^+$. If $\gamma_1\gamma_2 < 1$, then
(i)

$$\begin{aligned} \|e_{1t}\| &\leq (1-\gamma_1\gamma_2)^{-1}(\|u_{1t}\| + \gamma_2\|u_{2t}\| + \beta_2 + \gamma_2\beta_1) \\ \|e_{2t}\| &\leq (1-\gamma_1\gamma_2)^{-1}(\|u_{2t}\| + \gamma_1\|u_{1t}\| + \beta_1 + \gamma_1\beta_2) \end{aligned} \tag{2.27}$$

for any $t \geq 0$.
(ii) If in addition, $\|u_1\|, \|u_2\| < \infty$, then e_1, e_2, y_1, y_2 have finite norms, and the norms of e_1, e_2 are bounded by the right hand sides of (2.27) with all the subscripts of t dropped.

Proof. From Fig. 2.2, we have

$$e_1 = u_1 - y_2 = u_1 - H_2e_2 \tag{2.28}$$

$$e_2 = y_1 + u_2 = H_1e_1 + u_2 \tag{2.29}$$

$$\text{Thus } e_{1t} = u_{1t} - (H_2e_2)_t \tag{2.30}$$

$$e_{2t} = (H_1e_1)_t + u_{2t} \tag{2.31}$$

$\forall\ t \geq 0$. Taking norms on both sides of (2.30) and (2.31) and using the triangle inequality, we obtain

$$\begin{aligned} \|e_{1t}\| &\leq \|u_{1t}\| + \|(H_2e_2)_t\| \\ &\leq \|u_{1t}\| + \gamma_2\|e_{2t}\| + \beta_2 \\ &\quad \text{(using (2.26))} \end{aligned} \tag{2.32}$$

$$\begin{aligned} \|e_{2t}\| &\leq \|(H_1e_1)_t\| + \|u_{2t}\| \\ &\leq \gamma_1\|e_{1t}\| + \beta_1 + \|u_{2t}\| \\ &\quad \text{(using (2.25))} \end{aligned} \tag{2.33}$$

Now using (2.33) in (2.32) and rearranging terms we obtain

$$\|e_{1t}\| \leq (1-\gamma_1\gamma_2)^{-1}(\|u_{1t}\| + \gamma_2\|u_{2t}\| + \gamma_2\beta_1 + \beta_2)$$

which completes the proof of the first inequality in (2.27). The proof of the second inequality in (2.27) can be similarly obtained. The proof of (ii) follows immediately by using the fact that when $\|u_1\|, \|u_2\| < \infty$, then the right hand sides of (2.27) can be bounded using a constant that is independent of t.

The small gain theorem and its variations can be used to study the stability of a nominally stable system under perturbations. The following example shows the steps involved.

Example 2.3.1. Consider the configuration shown in Fig. 2.3. Here $P_0(s)$ and $Q(s)$ are stable proper rational transfer functions and $\Delta_m(s)$ is a stable transfer function such that $P_0(s)\Delta_m(s)$ is strictly proper. We would like to determine conditions under which $r \in L_2$ implies that all the other signals are also

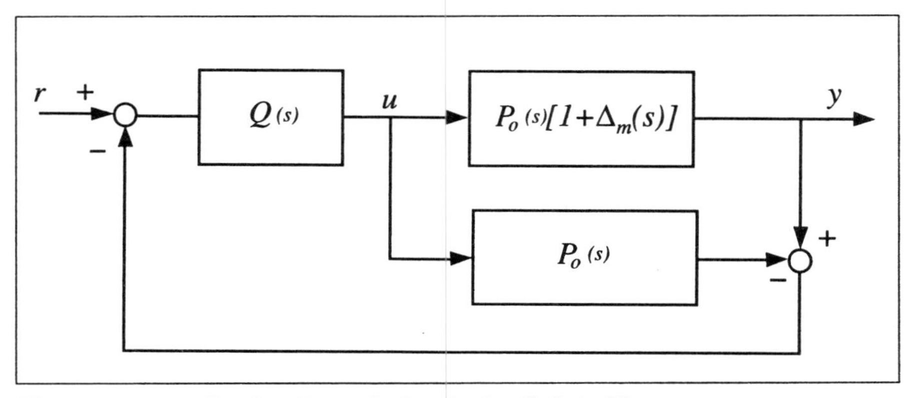

Fig. 2.3. An Application Example for the Small Gain Theorem

in L_2. Clearly this can be achieved by ensuring that $r \in L_2 \Rightarrow u \in L_2$. Now from Fig. 2.3, we obtain

$$\begin{aligned} u &= Q(s)[r - \{P_0(s)[1 + \Delta_m(s)][u] - P_0(s)[u]\}] \\ &= Q(s)[r] - P_0(s)\Delta_m(s)Q(s)[u] \\ \text{or } u_t &= [Q(s)[r]]_t - (P_0(s)\Delta_m(s)Q(s)[u])_t \\ \Rightarrow \|u_t\|_2 &\leq \|(P_0(s)\Delta_m(s)Q(s)[u])_t\|_2 + \|(Q(s)[r])_t\|_2 \\ &\leq \|P_0(s)\Delta_m(s)Q(s)\|_\infty \cdot \|u_t\|_2 + \|Q(s)\|_\infty \|r_t\|_2 \\ &\quad \text{(using Lemma 2.3.1 and Remark 2.3.2)} \end{aligned}$$

This shows that if

$$\|P_0(s)\Delta_m(s)Q(s)\|_\infty < 1 \tag{2.34}$$

then

$$\|u_t\|_2 \leq [1 - \|P_0(s)\Delta_m(s)Q(s)\|_\infty]^{-1}\|Q(s)\|_\infty \|r_t\|_2$$

so that $r \in L_2 \Rightarrow u \in L_2$. Thus (2.34) is a sufficient condition for the L_2 stability of the configuration shown in Fig. 2.3. Using some additional arguments [9], it can be shown that when $\Delta_m(s)$ is rational, (2.34) is also sufficient for L_∞ stability.

2.3.4 Bellman-Gronwall Lemma

A key lemma for the analysis of adaptive control schemes is the following.

Lemma 2.3.5. *(Bellman-Gronwall Lemma) [44] Let $\lambda(t), g(t), k(t)$ be non-negative piecewise continuous functions of time t. If the function $y(t)$ satisfies the inequality*

$$y(t) \leq \lambda(t) + g(t)\int_{t_0}^{t} k(s)y(s)ds, \ \forall\, t \geq t_0 \geq 0 \tag{2.35}$$

then

$$y(t) \le \lambda(t) + g(t)\int_{t_0}^{t} \lambda(s)k(s)[exp(\int_{s}^{t} k(\tau)g(\tau)d\tau)]ds$$
$$\forall\, t \ge t_0 \ge 0 \tag{2.36}$$

Proof. Let us define

$$q(t) \triangleq k(t)e^{-\int_{t_0}^{t} g(\tau)k(\tau)d\tau}$$

Because $k(t)$ is nonnegative, we have $q(t) \ge 0\ \forall\ t \ge t_0$. Multiplying both sides of (2.35) by $q(t)$, and rearranging the inequality, we obtain

$$q(t)y(t) - q(t)g(t)\int_{t_0}^{t} k(s)y(s)ds \le \lambda(t)q(t) \tag{2.37}$$

From the expression for $q(t)$, one can verify that

$$q(t)y(t) - q(t)g(t)\int_{t_0}^{t} k(s)y(s)ds = \frac{d}{dt}\left(e^{-\int_{t_0}^{t} g(\tau)k(\tau)d\tau}\int_{t_0}^{t} k(s)y(s)ds\right) \tag{2.38}$$

Using (2.38) in (2.37) and integrating both sides of (2.37), we obtain

$$e^{-\int_{t_0}^{t} g(\tau)k(\tau)d\tau}\int_{t_0}^{t} k(s)y(s)ds \le \int_{t_0}^{t} \lambda(s)q(s)ds$$

Therefore,

$$\begin{aligned}\int_{t_0}^{t} k(s)y(s)ds &\le e^{\int_{t_0}^{t} g(\tau)k(\tau)d\tau}\int_{t_0}^{t} \lambda(s)q(s)ds \\ &= e^{\int_{t_0}^{t} g(\tau)k(\tau)d\tau}\int_{t_0}^{t} \lambda(s)k(s)e^{-\int_{t_0}^{s} g(\tau)k(\tau)d\tau}ds \\ &= \int_{t_0}^{t} \lambda(s)k(s)e^{\int_{s}^{t} g(\tau)k(\tau)d\tau}ds \end{aligned} \tag{2.39}$$

Using (2.39) in (2.35), the proof for inequality (2.36) is complete.

2.4 Lyapunov Stability

2.4.1 Definition of Stability

We consider systems described by ordinary differential equations of the form

$$\dot{x} = f(t,x),\ \ x(t_0) = x_0 \tag{2.40}$$

where $x \in R^n$, $f : \mathcal{T} \times \mathcal{B}(r) \mapsto R$, $\mathcal{T} = [t_0, \infty)$ and $\mathcal{B}(r) = \{x \in R^n \,|\, |x| < r\}$. We assume that f is of such nature that for every $x_0 \in \mathcal{B}(r)$ and every $t_0 \in R^+$, (2.40) possesses one and only one solution $x(t, t_0, x_0)$.

Definition 2.4.1. *A state x_e is said to be an* equilibrium state *of the system described by (2.40) if $f(t, x_e) \equiv 0$ for all $t \geq t_0$.*

Definition 2.4.2. *An equilibrium state x_e is called an* isolated equilibrium state *if there exists a constant $r > 0$ such that $\mathcal{B}(x_e, r) \triangleq \{x | |x - x_e| < r\} \subset R^n$ contains no equilibrium state of (2.40) other than x_e.*

The equilibrium state $x_{1e} = 0$, $x_{2e} = 0$ of

$$\dot{x}_1 = x_1 x_2, \; \dot{x}_2 = x_1^2$$

is not isolated because any point $x_1 = 0, x_2 = \text{const.}$ is an equilibrium state. On the other hand, the differential equation

$$\dot{x} = (x-1)^2 x$$

has two isolated equilibrium states $x_e = 1$ and $x_e = 0$.

Definition 2.4.3. *The equilibrium state x_e is said to be* stable *(in the sense of Lyapunov) if for arbitrary t_0 and $\epsilon > 0$, there exists a $\delta(\epsilon, t_0)$ such that $|x_0 - x_e| < \delta \Rightarrow |x(t, t_0, x_0) - x_e| < \epsilon \, \forall \, t \geq t_0$.*

Definition 2.4.4. *The equilibrium state x_e is said to be* uniformly stable *(u.s.) if it is stable and if $\delta(\epsilon, t_0)$ in Definition 2.4.3 does not depend on t_0.*

Definition 2.4.5. *The equilibrium state x_e is said to be* asymptotically stable *(a.s.) if (i) it is stable and (ii) there exists a $\delta(t_0)$ such that $|x_0 - x_e| < \delta(t_0)$ implies $\lim_{t\to\infty} |x(t, t_0, x_0) - x_e| = 0$.*

Definition 2.4.6. *The set of all $x_0 \in R^n$ such that $x(t, t_0, x_0) \to x_e$ as $t \to \infty$ for some $t_0 \geq 0$ is called the* region of attraction *of the equilibrium state x_e. If condition (ii) of Definition 2.4.5 is satisfied, then the equilibrium state x_e is said to be* attractive.

Definition 2.4.7. *The equilibrium state x_e is said to be* uniformly asymptotically stable *(u.a.s.) if (i) it is uniformly stable and (ii) $\forall \, \epsilon > 0$ and any $t_0 \in R^+$, $\exists \, \delta_0$ (independent of t_0 and ϵ) and a $T(\epsilon) > 0$ (independent of t_0) such that $|x(t, t_0, x_0) - x_e| < \epsilon$ for all $t \geq t_0 + T(\epsilon)$ whenever $|x_0 - x_e| < \delta_0$.*

Definition 2.4.8. *The equilibrium state x_e is said to be* exponentially stable *(e.s.) if there exists an $\alpha > 0$, and $\forall \, \epsilon > 0$, $\exists \, \delta(\epsilon) > 0$ such that*

$$|x(t, t_0, x_0) - x_e| \leq \epsilon e^{-\alpha(t-t_0)} \text{ for all } t \geq t_0$$

whenever $|x_0 - x_e| < \delta(\epsilon)$.

Definition 2.4.9. *The equilibrium state x_e is said to be* unstable *if it is not stable.*

When (2.40) possesses a unique solution for each $x_0 \in R^n$ and $t_0 \in R^+$, we need the following definitions for the global characterization of solutions.

Definition 2.4.10. *A solution $x(t, t_0, x_0)$ of (2.40) is* bounded *if there exists a $\beta > 0$ such that $|x(t, t_0, x_0)| < \beta$ for all $t \geq t_0$, where β may depend on each solution.*

Definition 2.4.11. *The solutions of (2.40) are* uniformly bounded *(u.b.) if for any $\alpha > 0$ and $t_0 \in R^+$, there exists a $\beta = \beta(\alpha)$ independent of t_0 such that if $|x_0| < \alpha$, then $|x(t, t_0, x_0)| < \beta$ for all $t \geq t_0$.*

Definition 2.4.12. *The solutions of (2.40) are* uniformly ultimately bounded *(u.u.b.) (with bound B) if there exists a $B > 0$ and if corresponding to any $\alpha > 0$ and $t_0 \in R^+$, there exists a $T = T(\alpha) > 0$ (independent of t_0) such that $|x_0| < \alpha$ implies $|x(t, t_0, x_0)| < B$ for all $t \geq t_0 + T$*

Definition 2.4.13. *The equilibrium point x_e of (2.40) is* asymptotically stable in the large *(a.s. in the large) if it is stable and every solution of (2.40) tends to x_e as $t \to \infty$ (i.e. the region of attraction of x_e is all of R^n).*

Definition 2.4.14. *The equilibrium point x_e of (2.40) is* uniformly asymptotically stable in the large *(u.a.s. in the large) if (i) it is uniformly stable, (ii) the solutions of (2.40) are uniformly bounded, and (iii) for any $\alpha > 0$, any $\epsilon > 0$ and $t_0 \in R^+$, there exists $T(\epsilon, \alpha) > 0$ independent of t_0 such that if $|x_0 - x_e| < \alpha$ then $|x(t, t_0, x_0) - x_e| < \epsilon \; \forall \, t \geq t_0 + T(\epsilon, \alpha)$.*

Definition 2.4.15. *The equilibrium point x_e of (2.40) is* exponentially stable in the large *(e.s. in the large) if there exists $\alpha > 0$ and for any $\beta > 0$, there exists $k(\beta) > 0$ such that*

$$|x(t, t_0, x_0)| \leq k(\beta) e^{-\alpha(t-t_0)} \text{ for all } t \geq t_0$$

whenever $|x_0| < \beta$.

Definition 2.4.16. *If $x(t, t_0, x_0)$ is a solution of $\dot{x} = f(t, x)$, then the trajectory $x(t, t_0, x_0)$ is said to be* stable *(u.s., a.s., u.a.s., e.s., unstable) if the equilibrium point $z_e = 0$ of the differential equation*

$$\dot{z} = f(t, z + x(t, t_0, x_0)) - f(t, x(t, t_0, x_0))$$

is stable *(u.s., a.s., u.a.s., e.s., unstable respectively).*

The above stability concepts and definitions are illustrated by the following example:

Example 2.4.1. (i) $\dot{x} = 0$ has the equilibrium state $x_e = c$, where c is any constant. Clearly, this is not an isolated equilibrium state. It can be easily verified that $x_e = c$ is stable, u.s. but not a.s.
(ii) $\dot{x} = -x$ has an isolated equilibrium state $x_e = 0$. Its solution is given by

$x(t) = x(t, t_0, x_0) = e^{-(t-t_0)}x_0$. Now given any $\epsilon > 0$ choose $\delta := \epsilon$. Then $|x_0| < \delta$ implies that

$$|x(t)| = |e^{-(t-t_0)}x_0| \leq |x_0| < \epsilon \; \forall \, t \geq t_0 \geq 0 \tag{2.41}$$

Hence, according to Definition 2.4.3, $x_e = 0$ is stable. Because $\delta = \epsilon$ is independent of t_0, $x_e = 0$ is also u.s. Furthermore, it can be easily verified that $x_e = 0$ is a.s., u.a.s. and e.s. in the large.
(iii) $\dot{x} = -x^3$ has an isolated equilibrium state $x_e = 0$. Its solution is given by

$$x(t) = x(t, t_0, x_0) = \left(\frac{x_0^2}{1 + 2x_0^2(t - t_0)}\right)^{\frac{1}{2}}$$

Now given any $\epsilon > 0$, choose $\delta := \epsilon$. Then $|x_0| < \delta$ implies that

$$|x(t)| = \sqrt{\frac{x_0^2}{1 + 2x_0^2(t - t_0)}} \leq |x_0| < \epsilon \; \forall \, t \geq t_0 \geq 0 \tag{2.42}$$

Hence, according to Definition 2.4.3, $x_e = 0$ is stable. Because $\delta = \epsilon$ is independent of t_0, $x_e = 0$ is also u.s. Furthermore, because $x_e = 0$ is stable and $x(t) \to x_e = 0$ as $t \to \infty$ for all $x_0 \in R$, we have a.s. in the large. Let us now check whether $x_e = 0$ is u.a.s. in the large by using Definition 2.4.14. We have already shown u.s. From (2.42) we conclude that $x(t)$ is u.b. To satisfy condition (iii) of Definition 2.4.14, we need to find a $T > 0$ independent of t_0 such that for any $\alpha > 0$ and $\epsilon > 0$, $|x_0| < \alpha$ implies $|x(t)| < \epsilon$ for all $t \geq t_0 + T$. From (2.42), we have

$$\begin{aligned} |x(t)| &= \sqrt{\frac{x_0^2}{1 + 2x_0^2(t - t_0)}} \\ \Rightarrow |x(t)| &< \sqrt{\frac{1}{2T}} \; \forall \, t \geq t_0 + T \end{aligned}$$

Choosing $T = \frac{1}{2\epsilon^2}$, it follows that $|x(t)| < \epsilon \; \forall \; t \geq t_0 + T$. Hence $x_e = 0$ is u.a.s. in the large. Using Definition 2.4.15, we can conclude that $x_e = 0$ is not e.s.

2.4.2 Lyapunov's Direct Method

The stability properties of the equilibrium state or solution of (2.40) can be studied by using the so called direct method of Lyapunov (also known as Lyapunov's second method) [25, 26]. The objective of this method is to answer questions of stability by using the form of $f(t, x)$ in (2.40) rather than the explicit knowledge of the solutions. We start with the following definitions [28].

Definition 2.4.17. *A continuous function $\phi : [0,r] \mapsto R^+$ (or a continuous function $\phi : [0,\infty) \mapsto R^+$) is said to belong to* class $\mathcal{K}$ *i.e.* $\phi \in \mathcal{K}$ *if*
(i) $\phi(0) = 0$
(ii) ϕ *is strictly increasing on* $[0,r]$ *(or on* $[0,\infty)$*).*

Definition 2.4.18. *A continuous function $\phi : [0,\infty) \mapsto R^+$ is said to belong to* class $\mathcal{KR}$, *i.e.* $\phi \in \mathcal{KR}$ *if*
(i) $\phi(0) = 0$
(ii) ϕ *is strictly increasing on* $[0,\infty)$
(iii) $\lim_{r\to\infty} \phi(r) = \infty$.

The function $\phi(|x|) = \frac{x^2}{1+x^2}$ belongs to class $\mathcal{K}$ on $[0,\infty)$ but not to class $\mathcal{KR}$. The function $\phi(|x|) = |x|$ belongs to class $\mathcal{K}$ and class $\mathcal{KR}$.

Definition 2.4.19. *Two functions $\phi_1, \phi_2 \in \mathcal{K}$ defined on $[0,r]$ (or $[0,\infty)$) are said to be* of the same order of magnitude *if there exist positive constants k_1, k_2 such that*

$$k_1\phi_1(r_1) \le \phi_2(r_1) \le k_2\phi_1(r_1)\ \forall\ r_1 \in [0,r]\ (or\ \forall\ r_1 \in [0,\infty))$$

The functions $\phi_1(|x|) = \frac{x^2}{1+2x^2}$ and $\phi_2(|x|) = \frac{x^2}{1+x^2}$ are of the same order of magnitude since

$$\phi_1(|x|) \le \phi_2(|x|) = \frac{x^2}{1+x^2} = \frac{2x^2}{2+2x^2} \le \frac{2x^2}{1+2x^2} = 2\phi_1(|x|)$$

Definition 2.4.20. *A function $V(t,x) : R^+ \times \mathcal{B}(r) \mapsto R$ with $V(t,0) = 0\ \forall\ t \in R^+$ is* positive definite *if there exists a continuous function $\phi \in \mathcal{K}$ such that $V(t,x) \ge \phi(|x|)\ \forall\ t \in R^+,\ x \in \mathcal{B}(r)$ and some $r > 0$. $V(t,x)$ is called* negative definite *if $-V(t,x)$ is positive definite.*

The function $V(t,x) = \frac{x^2}{1-x^2}$ with $x \in \mathcal{B}(1)$ is positive definite, whereas $V(t,x) = \frac{1}{1+t}x^2$ is not. The function $V(t,x) = \frac{x^2}{1+x^2}$ is positive definite for all $x \in R$.

Definition 2.4.21. *A function $V(t,x) : R^+ \times \mathcal{B}(r) \mapsto R$ with $V(t,0) = 0\ \forall\ t \in R^+$ is said to be* positive (negative) semidefinite *if $V(t,x) \ge 0$ ($V(t,x) \le 0$) for all $t \in R^+$ and $x \in \mathcal{B}(r)$ for some $r > 0$.*

Definition 2.4.22. *A function $V(t,x) : R^+ \times \mathcal{B}(r) \mapsto R$ with $V(t,0) = 0\ \forall\ t \in R^+$ is said to be* decrescent *if there exists $\phi \in \mathcal{K}$ such that $|V(t,x)| \le \phi(|x|)\ \forall\ t \ge 0$ and $\forall\ x \in \mathcal{B}(r)$ for some $r > 0$.*

The function $V(t,x) = \frac{1}{1+t}x^2$ is decrescent because $V(t,x) = \frac{1}{1+t}x^2 \le x^2\ \forall\ t \in R^+$ but $V(t,x) = tx^2$ is not.

Definition 2.4.23. *A function $V(t,x) : R^+ \times R^n \mapsto R$ with $V(t,0) = 0 \; \forall \; t \in R^+$ is said to be* radially unbounded *if there exists $\phi \in \mathcal{KR}$ such that $V(t,x) \geq \phi(|x|)$ for all $x \in R^n$ and $t \in R^+$.*

The function $V(x) = \frac{x^2}{1+x^2}$ is positive definite (choose $\phi(|x|) = \frac{|x|^2}{1+|x|^2}$) but not radially unbounded. It is clear from Definition 2.4.23 that if $V(t,x)$ is radially unbounded, it is also positive definite for all $x \in R^n$ but the converse is not true. The reader should also be aware that in some textbooks "positive definite" is used for radially unbounded functions, and "locally positive definite" is used for our definition of positive definite functions.

Let us assume (without loss of generality) that $x_e = 0$ is an equilibrium point of (2.40) and define $\dot{V}$ to be the time derivative of the function $V(t,x)$ along the solution of (2.40) i.e.

$$\dot{V} = \frac{\partial V}{\partial t} + (\nabla V)^T f(t,x) \tag{2.43}$$

where $\nabla V = [\frac{\partial V}{\partial x_1}, \frac{\partial V}{\partial x_2}, \cdots, \frac{\partial V}{\partial x_n}]^T$ is the gradient of V with respect to x. The second method of Lyapunov is summarized by the following theorem.

Theorem 2.4.1. *Suppose there exists a positive definite function $V(t,x) : R^+ \times \mathcal{B}(r) \mapsto R$ for some $r > 0$ with continuous first-order partial derivatives with respect to x, t and $V(t,0) = 0 \; \forall \; t \in R^+$. Then the following statements are true:*
(i) If $\dot{V} \leq 0$, then $x_e = 0$ is stable.
(ii) If V is decrescent and $\dot{V} \leq 0$, then $x_e = 0$ is u.s.
(iii) If V is decrescent and $\dot{V} < 0$, then $x_e = 0$ is u.a.s.
(iv) If V is decrescent and there exist $\phi_1, \phi_2, \phi_3 \in \mathcal{K}$ of the same order of magnitude such that

$$\phi_1(|x|) \leq V(t,x) \leq \phi_2(|x|), \; \dot{V}(t,x) \leq -\phi_3(|x|)$$

for all $x \in \mathcal{B}(r)$ and $t \in R^+$, then $x_e = 0$ is e.s.

In the above theorem, the state x is restricted to be inside the ball $\mathcal{B}(r)$ for some $r > 0$. Therefore, the results (i) to (iv) of Theorem 2.4.1 are referred to as local results. Statement (iii) is equivalent to that there exist $\phi_1, \phi_2, \phi_3 \in \mathcal{K}$, where ϕ_1, ϕ_2, ϕ_3 do *not* have to be of the same order of magnitude, such that

$$\phi_1(|x|) \leq V(t,x) \leq \phi_2(|x|), \; \dot{V}(t,x) \leq -\phi_3(|x|).$$

Theorem 2.4.2. *Assume that (2.40) possesses unique solutions for all $x_0 \in R^n$. Suppose there exists a positive definite, decrescent and radially unbounded function $V(t,x) : R^+ \times R^n \mapsto R^+$ with continuous first order partial derivatives with respect to t, x and $V(t,0) = 0 \; \forall \; t \in R^+$. Then the following statements are true:*
(i) If $\dot{V} < 0$, then $x_e = 0$ is u.a.s. in the large.
(ii) If there exist $\phi_1, \phi_2, \phi_3 \in \mathcal{KR}$ of the same order of magnitude such that

$$\phi_1(|x|) \leq V(t,x) \leq \phi_2(|x|), \; \dot{V}(t,x) \leq -\phi_3(|x|)$$

then $x_e = 0$ *is* e.s. in the large.

Statement (i) of Theorem 2.4.2 is also equivalent to that there exist $\phi_1, \phi_2, \phi_3 \in \mathcal{KR}$ such that

$$\phi_1(|x|) \leq V(t,x) \leq \phi_2(|x|), \; \dot{V}(t,x) \leq -\phi_3(|x|), \; \forall \, x \in R^n.$$

Theorem 2.4.3. *Assume that (2.40) possesses unique solutions for all* $x_0 \in R^n$. *If there exists a function* $V(t,x)$ *defined on* $|x| \geq R$ *(where* R *may be large) and* $t \in [0,\infty)$ *with continuous first-order partial derivatives with respect to* x, t *and if there exist* $\phi_1, \phi_2 \in \mathcal{KR}$ *such that*
(i) $\phi_1(|x|) \leq V(t,x) \leq \phi_2(|x|)$
(ii) $\dot{V}(t,x) \leq 0$ *for all* $|x| \geq R$ *and* $t \in [0,\infty)$, *then the solutions of (2.40) are u.b. If, in addition, there exists* $\phi_3 \in \mathcal{K}$ *defined on* $[0,\infty)$ *and*
(iii) $\dot{V}(t,x) \leq -\phi_3(|x|)$ *for all* $|x| \geq R$ *and* $t \in [0,\infty)$, *then the solutions of (2.40) are u.u.b.*

The system (2.40) is referred to as *nonautonomous*. When the function f in (2.40) does not depend explicitly on time t, the system is referred to as *autonomous*. In this case, we write

$$\dot{x} = f(x), \; x(t_0) = x_0 \tag{2.44}$$

Since (2.44) is a special case of (2.40), all the earlier theorems apply. However, the words "decrescent" and "uniform" can be dropped here since $V(t,x) = V(x)$, (i.e. does not depend on t) and the right hand side of (2.44) being independent of t, the stability (a.s.) is always uniform.

For the system (2.44), we can obtain a stronger result than Theorem 2.4.2 for a.s. as indicated below. To do so, we need the following definition.

Definition 2.4.24. *A set* Ω *in* R^n *is* invariant *with respect to the equation (2.44) if every solution of (2.44) starting in* Ω *remains in* Ω *for all* t.

Theorem 2.4.4. *(Lasalle Invariance Principle) Assume that (2.44) possesses unique solutions for all* $x_0 \in R^n$. *Suppose there exists a positive definite and radially unbounded function* $V(x) : R^n \mapsto R^+$ *with continuous first-order derivative with respect to* x *and* $V(0) = 0$. *If*
(i) $\dot{V} \leq 0 \, \forall \, x \in R^n$
(ii) The origin $x = 0$ *is the only invariant subset of the set* $\Omega = \{x \in R^n | \dot{V} = 0\}$, *then the equilibrium* $x_e = 0$ *of (2.44) is a.s. in the large*

Theorems 2.4.1 to 2.4.4 are referred to as Lyapunov-type theorems. The function $V(t,x)$ or $V(x)$ that satisfies any Lyapunov-type theorem is referred to as a Lyapunov function.

The following examples demonstrate the use of Lyapunov's direct method to analyze the stability of nonlinear systems.

Example 2.4.2. Consider the nonlinear system

$$\begin{aligned} \dot{x}_1 &= x_2 + cx_1(x_1^2 + x_2^2) \\ \dot{x}_2 &= -x_1 + cx_2(x_1^2 + x_2^2) \end{aligned} \tag{2.45}$$

where c is a constant. Note that $x_e = 0$ is the only equilibrium state. Let us choose

$$V(x) = x_1^2 + x_2^2$$

as a candidate for a Lyapunov function. Clearly $V(x)$ is positive definite, decrescent and radially unbounded. Furthermore, its time derivative along the solution of (2.45) is

$$\dot{V} = 2c(x_1^2 + x_2^2)^2 \tag{2.46}$$

If $c = 0$, then $\dot{V} = 0$ and, therefore, $x_e = 0$ is u.s. If $c < 0$, then $\dot{V} = -2|c|(x_1^2 + x_2^2)^2$ is negative definite, and therefore, $x_e = 0$ is u.a.s. in the large. If $c > 0$, then we can show that $x_e = 0$ is unstable. This conclusion, however, does not follow from any of the theorems presented so far; instead, instability theorems based on the second method of Lyapunov [44] have to be invoked.

Example 2.4.3. Consider the following system describing the motion of a simple pendulum

$$\begin{aligned} \dot{x}_1 &= x_2 \\ \dot{x}_2 &= -k \sin x_1 \end{aligned} \tag{2.47}$$

where $k > 0$ is a constant, x_1 is the angle, and x_2 the angular velocity. We consider a candidate for a Lyapunov function, the function $V(x)$ representing the total energy of the pendulum given as the sum of the kinetic and potential energies, i.e.

$$V(x) = \frac{1}{2}x_2^2 + k\int_0^{x_1} \sin \eta d\eta = \frac{1}{2}x_2^2 + k(1 - \cos x_1)$$

This $V(x)$ is positive definite and decrescent $\forall\, x \in \mathcal{B}(r)$ for some $r \in (0,\ \pi)$, but not radially unbounded. Along the solution of (2.47), we have

$$\dot{V} = 0$$

Therefore, the equilibrium state $x_e = 0$ is u.s.

Example 2.4.4. Consider the system

$$\begin{aligned} \dot{x}_1 &= x_2 \\ \dot{x}_2 &= -x_2 - e^{-t}x_1 \end{aligned} \tag{2.48}$$

Let us choose the positive definite, decrescent, and radially unbounded function

$$V(x) = x_1^2 + x_2^2$$

as a Lyapunov function candidate. Then along the solution of (2.48), we have

$$\dot{V} = -2x_2^2 + 2x_1x_2(1 - e^{-t})$$

which is sign indefinite so that none of the preceding Lyapunov theorems is applicable and no conclusion can be reached. Consequently, let us try another V function

$$V(t, x) = x_1^2 + e^t x_2^2$$

In this case, we obtain

$$\dot{V}(t, x) = -e^t x_2^2$$

This V function is positive definite, and $\dot{V}$ is negative semidefinite. Therefore, Theorem 2.4.1 is applicable, and we conclude that the equilibrium state $x_e = 0$ is stable. However, since V is not decrescent, we cannot conclude that the equilibrium state $x_e = 0$ is u.s.

Example 2.4.5. Consider the differential equation

$$\begin{aligned} \dot{x}_1 &= -2x_1 + x_1x_2 + x_2 \\ \dot{x}_2 &= -x_1^2 - x_1 \end{aligned} \tag{2.49}$$

and the Lyapunov function candidate $V(x) = \frac{x_1^2}{2} + \frac{x_2^2}{2}$. Then along the solutions of (2.49) we have $\dot{V} = -2x_1^2 \leq 0$ and the equilibrium state $x_{1e} = 0, x_{2e} = 0$ is u.s. The set defined in Theorem 2.4.4 is given by

$$\Omega = \{(x_1, x_2) | x_1 = 0\}$$

Because $\dot{x}_1 = x_2$ on Ω, any solution that starts from Ω with $x_2 \neq 0$ leaves Ω. Hence, $x_1 = 0$, $x_2 = 0$ is the only invariant subset of Ω. Therefore, the equilibrium $x_{1e} = 0$, $x_{2e} = 0$ is a.s. in the large.

The main drawback of Lyapunov's direct method is that, in general, there is no procedure for finding the appropriate Lyapunov function that satisfies the conditions of Theorems 2.4.1 to 2.4.4 except in the case where (2.40) represents a linear time invariant system.

2.4.3 Lyapunov-Like Functions

The choice of an appropriate Lyapunov function for establishing stability using Theorems 2.4.1 to 2.4.4 may not be obvious or possible in the case of many adaptive systems. However, a function that resembles a Lyapunov function, but does not possess all the properties that are needed to apply Theorems 2.4.1 to 2.4.4, can quite often be used to study the stability and boundedness properties of adaptive systems. Such a function is referred to as a *Lyapunov-like function*. The following example illustrates the use of Lyapunov-like functions.

Example 2.4.6. Consider the third-order differential equation

$$\begin{aligned}\dot{x}_1 &= -x_1 - x_2x_3, \; x_1(0) = x_{10}\\ \dot{x}_2 &= x_1x_3, \; x_2(0) = x_{20}\\ \dot{x}_3 &= x_1^2, \; x_3(0) = x_{30}\end{aligned} \tag{2.50}$$

which has the nonisolated equilibrium points in R^3 defined by $x_1 = 0$, $x_2 =$ const., $x_3 = 0$ or $x_1 = 0$, $x_2 = 0$, $x_3 =$ const. We would like to analyze the stability properties of the solutions of (2.50) by using an appropriate Lyapunov function and applying Theorems 2.4.1 to 2.4.4. If we follow Theorems 2.4.1 to 2.4.4, then we should start with a function $V(x_1, x_2, x_3)$ that is positive definite in R^3. Instead of doing so, let us consider the simple quadratic function

$$V(x_1, x_2) = \frac{x_1^2}{2} + \frac{x_2^2}{2}$$

which is positive semidefinite in R^3 and, therefore, does not satisfy the positive definite condition of Theorems 2.4.1 to 2.4.4. The time derivative of V along the solution of the differential equation (2.50) satisfies

$$\dot{V} = -x_1^2 \leq 0 \tag{2.51}$$

which implies that V is a nonincreasing function of time. Therefore,

$$V(x_1(t), x_2(t)) \leq V(x_1(0), x_2(0)) \triangleq V_0$$

and $V, x_1, x_2 \in L_\infty$. Furthermore, by Lemma 2.2.5 (i), V has a limit as $t \to \infty$, i.e.

$$\lim_{t\to\infty} V(x_1(t), x_2(t)) = V_\infty$$

so that (2.51) implies that

$$\int_0^\infty x_1^2(\tau)d\tau = V_0 - V_\infty \; < \; \infty$$

i.e. $x_1 \in L_2$. Using $x_1 \in L_2$ in (2.50) we obtain $x_3 \in L_\infty$. Moreover, using $x_1, x_2, x_3 \in L_\infty$, we conclude from (2.50) that $\dot{x}_1 \in L_\infty$. Using $\dot{x}_1 \in L_\infty$, $x_1 \in L_2 \cap L_\infty$ and applying Corollary 2.2.1, we have $x_1(t) \to 0$ as $t \to \infty$. By using the properties of the positive semidefinite function $V(x_1, x_2)$, we have established that the solution of (2.50) is uniformly bounded and $x_1(t) \to 0$ as $t \to \infty$ for any finite initial condition $x_1(0)$, $x_2(0)$, $x_3(0)$. Because the approach we followed resembles the Lyapunov function approach, we are motivated to refer to $V(x_1, x_2)$ as a Lyapunov-like function.

We use Lyapunov-like functions and similar arguments as in the example above to analyze the stability of a wide class of adaptive schemes considered in this monograph.

2.4.4 Lyapunov's Indirect Method

Under certain conditions, conclusions can be drawn about the stability of the equilibrium of a nonlinear system by studying the behaviour of a certain linear system obtained by linearizing (2.40) around its equilibrium state. This method is known as the *first method of Lyapunov* or as *Lyapunov's indirect method* and is described as follows [2, 44]: Let $x_e = 0$ be an equilibrium state of (2.40) and assume that $f(t, x)$ is continuously differentiable with respect to x for each $t \geq 0$. Then in the neighbourhood of $x_e = 0$, f has a Taylor series expansion that can be written as

$$\dot{x} = f(t, x) = A(t)x + f_1(t, x) \tag{2.52}$$

where $A(t) = \nabla f|_{x=0}$ is referred to as the Jacobian matrix of f evaluated at $x = 0$ and $f_1(t, x)$ represents the remaining terms in the series expansion.

Theorem 2.4.5. *Assume that $A(t)$ is uniformly bounded and that*

$$\lim_{|x| \to 0} \sup_{t \geq 0} \frac{|f_1(t, x)|}{|x|} = 0$$

Let $z_e = 0$ be the equilibrium state of

$$\dot{z}(t) = A(t)z(t).$$

Then the following statements are true for the equilibrium $x_e = 0$ of (2.52):
(i) If $z_e = 0$ is u.a.s. then $x_e = 0$ is u.a.s.
(ii) If $z_e = 0$ is unstable then $x_e = 0$ is unstable
(iii) If $z_e = 0$ is u.s. or stable, no conclusions can be drawn about the stability of $x_e = 0$.

The proof of Theorem 2.4.5 can be found in [44].

2.4.5 Stability of Linear Systems

Equation (2.52) indicates that certain classes of nonlinear systems can be approximated by linear ones in the neighbourhood of an equilibrium point or, as often called in practice, an operating point. For this reason, it is important to study the stability of linear systems of the form

$$\dot{x}(t) = A(t)x(t) \tag{2.53}$$

where the elements of $A(t)$ are piecewise continuous for all $t \geq t_0 \geq 0$, as a special class of the nonlinear system (2.40) or as an approximation of the linearized system (2.52). The solution of (2.53) is given by [21]

$$x(t, t_0, x_0) = \Phi(t, t_0)x_0$$

for all $t \geq t_0$, where $\Phi(t, t_0)$ is the *state transition matrix* and satisfies the matrix differential equation

$$\begin{aligned}\frac{\partial}{\partial t}\Phi(t, t_0) &= A(t)\Phi(t, t_0), \ \forall\, t \geq t_0 \\ \Phi(t_0, t_0) &= I\end{aligned}$$

Some additional useful properties of $\Phi(t, t_0)$ are
(i) $\Phi(t, t_0) = \Phi(t, \tau)\Phi(\tau, t_0)\ \forall\, t \geq \tau \geq t_0$
(ii) $\Phi(t, t_0)^{-1} = \Phi(t_0, t)$
(iii) $\frac{\partial}{\partial t_0}\Phi(t, t_0) = -\Phi(t, t_0)A(t_0)$

Necessary and sufficient conditions for the stability of the equilibrium state $x_e = 0$ of (2.53) are given by the following theorem.

Theorem 2.4.6. *Let $||\Phi(t, \tau)||$ denote the induced matrix norm of $\Phi(t, \tau)$ at each time $t \geq \tau$. The equilibrium state $x_e = 0$ of (2.53) is*
(i) stable if and only if the solutions of (2.53) are bounded or equivalently

$$c(t_0) \triangleq \sup_{t \geq t_0} ||\Phi(t, t_0)|| < \infty$$

(ii) u.s. if and only if

$$c_0 \triangleq \sup_{t_0 \geq 0} c(t_0) = \sup_{t_0 \geq 0}\left(\sup_{t \geq t_0} ||\Phi(t, t_0)||\right) < \infty$$

(iii) a.s. if and only if

$$\lim_{t \to \infty} ||\Phi(t, t_0)|| = 0$$

for any $t_0 \in R^+$
(iv) u.a.s. if and only if there exist positive constants α and β such that

$$||\Phi(t, t_0)|| \leq \alpha e^{-\beta(t - t_0)}, \ \forall\, t \geq t_0 \geq 0$$

(v) e.s. if and only if it is u.a.s.
(vi) a.s., u.a.s, e.s. in the large if and only if it is a.s., u.a.s., e.s. respectively.

When $A(t) = A$ is a constant matrix, the conditions for stability of the equilibrium $x_e = 0$ of

$$\dot{x} = Ax \tag{2.54}$$

are given by the following theorems.

Theorem 2.4.7. *The equilibrium state $x_e = 0$ of (2.54) is stable if and only if*
(i) All the eigenvalues of A have nonpositive real parts
(ii) For each eigenvalue λ_i with $Re\{\lambda_i\} = 0$, λ_i is a simple zero of the minimal polynomial of A (i.e., of the monic polynomial $\psi(\lambda)$ of least degree such that $\psi(A) = 0$).

Theorem 2.4.8. *A necessary and sufficient condition for $x_e = 0$ to be a.s. in the large is that any one of the following conditions is satisfied:*
(i) All the eigenvalues of A have negative real parts
(ii) For every positive definite matrix Q, the following Lyapunov matrix equation

$$A^T P + PA = -Q$$

has a unique solution P that is also positive definite.

It can be easily verified that for the linear time-invariant system given by (2.54), if $x_e = 0$ is stable, it is also u.s. If $x_e = 0$ is a.s., it is also u.a.s and e.s. in the large.

In the rest of this monograph, we will abuse notation and say that the matrix A in (2.54) is *stable* when the equilibrium $x_e = 0$ is a.s., i.e. when all the eigenvalues of A have negative real parts and *marginally stable* when $x_e = 0$ is stable, i.e. A satisfies (i) and (ii) of Theorem 2.4.7.

Let us consider again the linear time-varying system (2.53) and suppose that for each fixed t all eigenvalues of the matrix $A(t)$ have negative real parts. In view of Theorem 2.4.8, one may ask whether this condition on $A(t)$ can ensure some form of stability for the equilibrium $x_e = 0$ of (2.53). The answer is unfortunately no in general, as demonstrated by the following example given in [44].

Example 2.4.7. Let

$$A(t) = \begin{bmatrix} -1 + 1.5\cos^2 t & 1 - 1.5\sin t \cos t \\ -1 - 1.5\sin t \cos t & -1 + 1.5\sin^2 t \end{bmatrix}$$

The eigenvalues of $A(t)$ for each fixed t,

$$\lambda(A(t)) = -0.25 \pm j0.5\sqrt{1.75}$$

have negative real parts and are also independent of t. Despite this the equilibrium $x_e = 0$ of (2.53) is unstable because

$$\Phi(t, 0) = \begin{bmatrix} e^{0.5t}\cos t & e^{-t}\sin t \\ -e^{0.5t}\sin t & e^{-t}\cos t \end{bmatrix}$$

is unbounded with respect to time t.

Despite Example 2.4.7, Theorem 2.4.8 can be used to obtain some sufficient conditions for $A(t)$, which guarantee that $x_e = 0$ of (2.53) is u.a.s. as indicated by the following theorem.

Theorem 2.4.9. *Let the elements of $A(t)$ in (2.53) be differentiable and bounded functions of time and assume that*
(A1) $Re\{\lambda_i(A(t))\} \le -\sigma_s$ $\forall\, t \ge 0$ and for $i = 1, 2, \cdots, n$ where $\sigma_s > 0$ is some constant.
(i) If $\|\dot{A}\| \in L_2$, then the equilibrium state $x_e = 0$ of (2.53) is u.a.s. in the

large.
(ii) If any one of the following conditions:
(a) $\int_t^{t+T} \|\dot{A}(\tau)\| d\tau \leq \mu T + \alpha_0$, *i.e.* $(\|\dot{A}\|)^{\frac{1}{2}} \in \mathcal{S}(\mu)$
(b) $\int_t^{t+T} \|\dot{A}(\tau)\|^2 d\tau \leq \mu^2 T + \alpha_0$, *i.e.* $\|\dot{A}\| \in \mathcal{S}(\mu^2)$
(c) $\|\dot{A}(t)\| \leq \mu$
is satisfied for some α_0, $\mu \in R^+$ *and* $\forall\, t, T \geq 0$, *then there exists a* $\mu^* > 0$ *such that if* $\mu \in [0, \mu^*)$, *the equilibrium state* $x_e = 0$ *of (2.53) is u.a.s. in the large.*

Proof. Using (A1), it follows from Theorem 2.4.8 that the Lyapunov equation

$$A^T(t)P(t) + P(t)A(t) = -I \tag{2.55}$$

has a unique bounded solution $P(t)$ for each fixed t. We consider the following Lyapunov function candidate:

$$V(t, x) = x^T P(t) x$$

Then along the solution of (2.53) we have

$$\dot{V} = -|x(t)|^2 + x^T(t)\dot{P}(t)x(t) \tag{2.56}$$

From (2.55), $\dot{P}$ satisfies

$$A^T(t)\dot{P}(t) + \dot{P}(t)A(t) = -Q(t),\ \forall\, t \geq 0 \tag{2.57}$$

where $Q(t) = \dot{A}^T(t)P(t) + P(t)\dot{A}(t)$. Because of (A1), it can be verified [21] that

$$\dot{P}(t) = \int_0^\infty e^{A^T(t)\tau} Q(t) e^{A(t)\tau} d\tau$$

satisfies (2.57) for each $t \geq 0$, so that

$$\|\dot{P}(t)\| \leq \|Q(t)\| \int_0^\infty \|e^{A^T(t)\tau}\|.\|e^{A(t)\tau}\| d\tau$$

Because (A1) implies that $\|e^{A(t)\tau}\| \leq \alpha_1 e^{-\alpha_0 \tau}$ for some $\alpha_1,\ \alpha_0 > 0$, it follows that

$$\|\dot{P}(t)\| \leq c\|Q(t)\|$$

for some $c \geq 0$. Then,

$$\|Q(t)\| \leq 2\|P(t)\|.\|\dot{A}(t)\| \text{ (by (2.57))}$$

together with $P \in L_\infty$ imply that

$$\|\dot{P}\| \leq \beta\|\dot{A}(t)\|\ \forall\, t \geq 0 \tag{2.58}$$

for some constant $\beta \geq 0$. Using (2.58) in (2.56) and noting that P satisfies $0 < \beta_1 \leq \lambda_{\min}(P) \leq \lambda_{\max}(P) \leq \beta_2$ for some $\beta_1, \beta_2 > 0$, we obtain

$$\dot{V}(t) \leq -|x(t)|^2 + \beta\|\dot{A}(t)\|.|x(t)|^2 \leq -\beta_2^{-1}V(t) + \beta\beta_1^{-1}\|\dot{A}(t)\|V(t)$$

so that

$$V(t) \leq e^{-\int_{t_0}^{t}(\beta_2^{-1} - \beta\beta_1^{-1}\|\dot{A}(\tau)\|)d\tau}V(t_0) \tag{2.59}$$

Let us prove (ii) first. Using condition (a) in (2.59) we have

$$V(t) \leq e^{-(\beta_2^{-1} - \beta\beta_1^{-1}\mu)(t-t_0)}e^{\beta\beta_1^{-1}\alpha_0}V(t_0)$$

Therefore, for $\mu^* = \frac{\beta_1}{\beta_2\beta}$ and $\forall\ \mu \in [0, \mu^*)$, $V(t) \to 0$ exponentially fast, which implies that $x_e = 0$ is u.a.s. in the large.

To be able to use (b), we rewrite (2.59) as

$$V(t) \leq e^{-\beta_2^{-1}(t-t_0)}e^{\beta\beta_1^{-1}\int_{t_0}^{t}\|\dot{A}(\tau)\|d\tau}V(t_0)$$

Using the Schwartz inequality and (b) we have

$$\begin{aligned}
\int_{t_0}^{t}\|\dot{A}(\tau)\|d\tau &\leq \left(\int_{t_0}^{t}\|\dot{A}(\tau)\|^2 d\tau\right)^{\frac{1}{2}}\sqrt{t-t_0} \\
&\leq [\mu^2(t-t_0)^2 + \alpha_0(t-t_0)]^{\frac{1}{2}} \\
&\leq \mu(t-t_0) + \sqrt{\alpha_0}\sqrt{t-t_0}
\end{aligned}$$

Therefore, $V(t) \leq e^{-\alpha(t-t_0)}y(t)V(t_0)$ where $\alpha = (1-\gamma)\beta_2^{-1} - \beta\beta_1^{-1}\mu$,

$$\begin{aligned}
y(t) &= \exp[-\gamma\beta_2^{-1}(t-t_0) + \beta\beta_1^{-1}\sqrt{\alpha_0}\sqrt{t-t_0}] \\
&= \exp\left[-\gamma\beta_2^{-1}\left(\sqrt{t-t_0} - \frac{\beta\beta_1^{-1}\sqrt{\alpha_0}}{2\gamma\beta_2^{-1}}\right)^2 + \frac{\alpha_0\beta^2\beta_2}{4\gamma\beta_1^2}\right]
\end{aligned}$$

and γ is an arbitrary constant that satisfies $0 < \gamma < 1$. It is clear from the above expression that

$$y(t) \leq \exp\left[\alpha_0\frac{\beta^2\beta_2}{4\gamma\beta_1^2}\right] \triangleq c\ \forall\ t \geq t_0$$

Hence $V(t) \leq ce^{-\alpha(t-t_0)}V(t_0)$. Choosing $\mu^* = \frac{\beta_1(1-\gamma)}{\beta_2\beta}$, it follows that $\forall\ \mu \in [0, \mu^*)$, $\alpha > 0$ and, therefore, $V(t) \to 0$ exponentially fast, which implies that $x_e = 0$ is u.a.s. in the large.

Since (c) implies (a), the proof of (c) follows directly from that of (a).

The proof of (i) follows from that of (ii) (b), because $\|\dot{A}\| \in L_2$ implies (b) with $\mu = 0$.

Theorem 2.4.9 essentially states that if the eigenvalues of $A(t)$ for each fixed t have negative real parts and if $A(t)$ varies sufficiently slowly most of the time, then the equilibrium state $x_e = 0$ of (2.53) is u.a.s.

CHAPTER 3
INTERNAL MODEL CONTROL SCHEMES

3.1 Introduction

In this chapter, we introduce the class of internal model control (IMC) schemes. These schemes derive their name from the fact that the controller implementation includes an explicit model of the plant as a part of the controller. Such schemes enjoy immense popularity in process control applications where, in most cases, the plant to be controlled is open-loop stable. As will be seen in this chapter, the IMC configuration for a stable plant is really a particular case of the Youla-Jabr-Bongiorno-Kucera (YJBK) parametrization of all controllers that preserve closed loop stability [46]. The IMC parameter here plays the same role as the Youla parameter in the YJBK parametrization, and consequently can be chosen to meet one of several design objectives. Choices of the IMC parameter that lead to some familiar control schemes are discussed. Finally, the chapter concludes with an appraisal of the inherent robustness of the IMC structure to plant perturbations. Throughout this chapter, as indeed in the rest of this monograph, we focus on plants whose modelled parts are linear, time-invariant and finite dimensional. The controllers considered are also assumed to belong to the same class so that the modelled part of the plant and the controller can both be described in the Laplace domain using rational, proper transfer functions.

3.2 The Internal Model Control Structure and the YJBK Parametrization

In this section, we first introduce the Internal Model Control Structure. To this end, we consider the IMC configuration for a stable plant $P(s)$ as shown in Figure 3.1. The IMC controller consists of a stable "IMC parameter" $Q(s)$ and a model of the plant which is usually referred to as the "internal model." It can be shown[30, 6] that if the plant $P(s)$ is stable and the internal model is an exact replica of the plant, then the stability of the IMC parameter is equivalent to the internal stability of the configuration in Figure 3.1. Indeed, the IMC parameter is really the Youla parameter[46] that appears in a special case of the YJBK parametrization of all stabilizing controllers [6]. Because

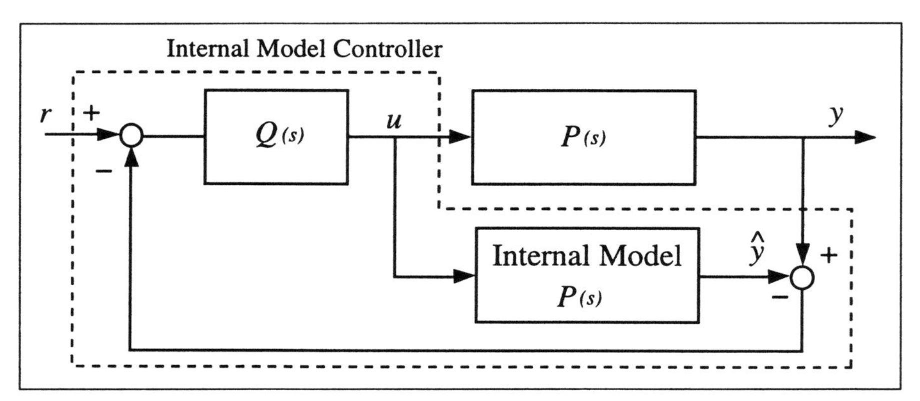

Fig. 3.1. The Internal Model Control Structure

of this, internal stability is assured as long as $Q(s)$ is chosen to be any stable rational transfer function.

We now derive the internal model control structure as a special case of the YJBK parametrization [46]. A detailed treatment of the YJBK parametrization is given in Appendix A.

Consider the standard unity feedback configuration shown in Figure 3.2. Here u and y represent the plant input and output signals respectively, r is

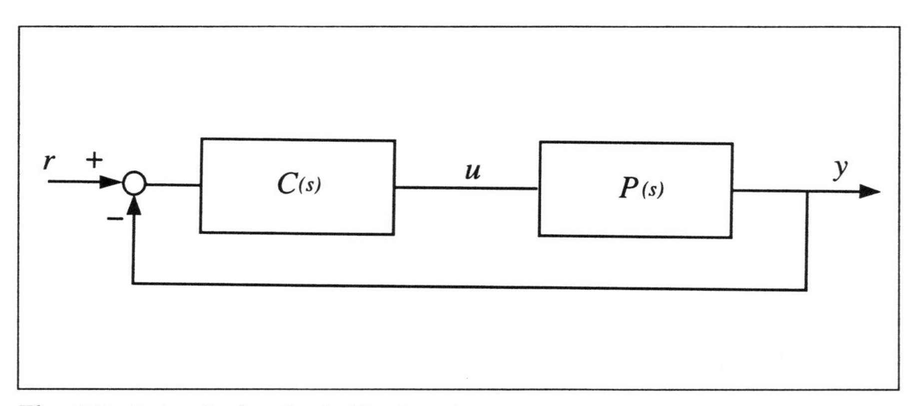

Fig. 3.2. A standard unity feedback system

the command signal, $P(s)$ and $C(s)$ are the rational proper transfer functions of the plant and the compensator respectively. The YJBK parametrization seeks to parametrize the set of *all* compensators $C(s)$ that guarantee the stability of the closed-loop system. To do so, we consider a coprime factorization $N_p(s)/D_p(s)$ of the plant $P(s)$, where $N_p(s)$ and $D_p(s)$ are stable rational transfer functions. Let $X(s)$ and $Y(s)$ denote a solution of the Bezout identity

$$X(s)N_p(s) + Y(s)D_p(s) = 1 \tag{3.1}$$

in the set of stable rational transfer functions. Then the set of all stabilizing controllers $C(s)$ is given by (see Appendix A)

$$C(s) = \frac{X(s) + D_p(s)Q(s)}{Y(s) - N_p(s)Q(s)} \tag{3.2}$$

where $Q(s)$ varies over the set of all stable rational transfer functions.

Now suppose the plant $P(s)$ is open-loop stable. Then a possible choice of coprime factors is $N_p(s) = P(s)$ and $D_p(s) = 1$, so that the Bezout identity (3.1) holds with $X(s) = 0$ and $Y(s) = 1$. Then it follows from (3.2) that every stabilizing compensator $C(s)$ is given by

$$C(s) = \frac{Q(s)}{1 - P(s)Q(s)}, \tag{3.3}$$

where $Q(s)$ varies over the set of all stable rational transfer functions. This compensator when inserted into the setup in Figure 3.2 leads to the Youla parametrized system shown in Figure 3.3. Keeping the external input-output

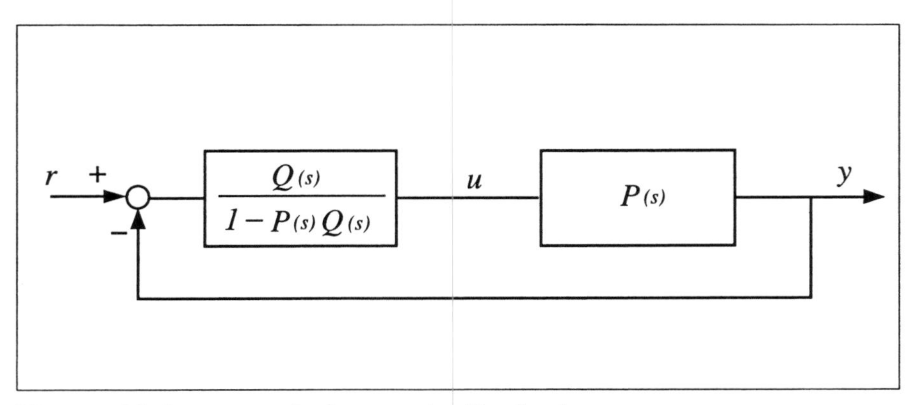

Fig. 3.3. Youla parametrized system (stable plant)

behaviour the same, the above system can be successively transformed into the configurations shown in Figures 3.4 and 3.5 respectively.

The configuration in Fig. 3.5 is referred to as the internal model control (IMC) structure, because the controller in this configuration incorporates an explicit model of the plant. Thus the IMC structure for stable plants is a special case of the YJBK parametrization of all stabilizing compensators. Since $Q(s)$ is the Youla parameter, it follows that the IMC configuration in Fig. 3.5 is stable if and only if $Q(s)$ is a stable transfer function.

From Fig. 3.5, we see that the control input u in the IMC structure is given by

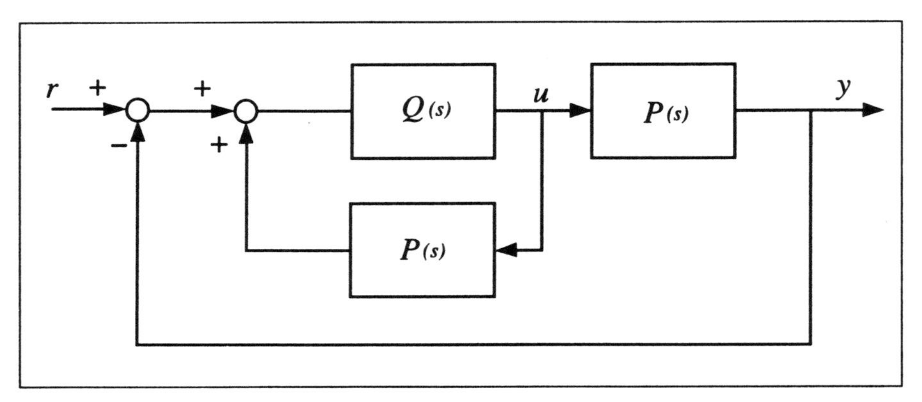

Fig. 3.4. Equivalent feedback system

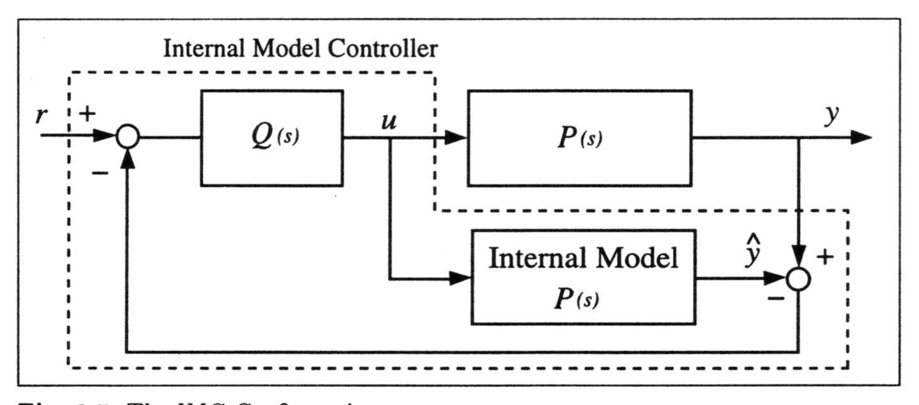

Fig. 3.5. The IMC Configuration.

$$u = Q(s)[r - (y - \hat{y})] \tag{3.4}$$

where $Q(s)$ is a stable rational transfer function. An alternative implementation of the control law (3.4) which is particularly well suited for adapting the coefficients of $Q(s)$ on-line can be derived as follows. Express $Q(s)$ as $Q(s) = \frac{Q_n(s)}{Q_d(s)}$ where $Q_n(s)$, $Q_d(s)$ are polynomials with $Q_d(s)$ being monic and of degree n_d. Let $\Lambda_1(s)$ be an arbitrary monic Hurwitz polynomial of degree n_d. Then the control law (3.4) can also be implemented as

$$u = q_d^T \frac{a_{n_d-1}(s)}{\Lambda_1(s)}[u] + q_n^T \frac{a_{n_d}(s)}{\Lambda_1(s)}[r - (y - \hat{y})] \tag{3.5}$$

where $\Lambda_1(s) - Q_d(s) = q_d^T a_{n_d-1}(s)$; $Q_n(s) = q_n^T a_{n_d}(s)$; $a_{n_d}(s) = [s^{n_d}, s^{n_d-1}, \cdots, 1]^T$; and $a_{n_d-1}(s) = [s^{n_d-1}, s^{n_d-2}, \cdots, 1]^T$. In other words, q_d and q_n are the vectors of coefficients of $\Lambda_1(s) - Q_d(s)$ and $Q_n(s)$ respectively. The implementation (3.5) of the IMC control law is depicted in Fig. 3.6 below.

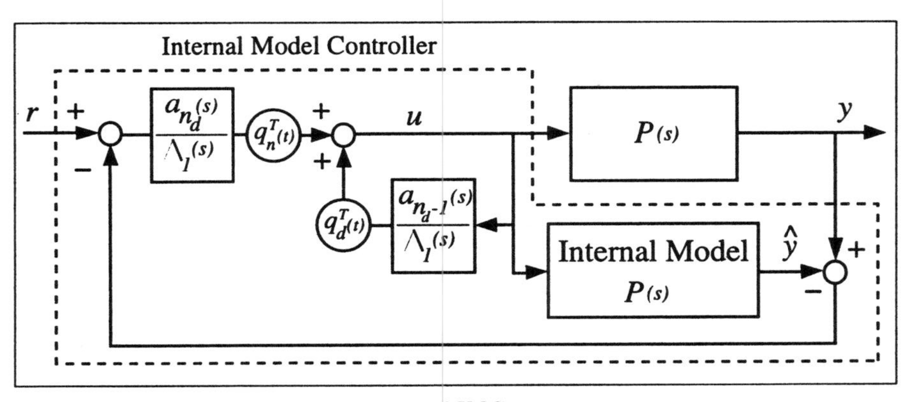

Fig. 3.6. Alternative Implementation of IMC

Remark 3.2.1. The implementation of the IMC structure as in Figs. 3.5 and 3.6 is appropriate only for open-loop stable plants. Indeed, an open loop unstable plant in these configurations will continue to be closed loop unstable, since perfect matching between the plant and the model will ensure that there is effectively no feedback signal for stabilization purposes. Nevertheless, the IMC structure can be used for design purposes even for open-loop unstable plants e.g. [30], and in this case the parameter $Q(s)$ corresponds to the Q parametrization [47, 48]. To guarantee closed-loop stability, $Q(s)$ must not only be stable but must also satisfy certain constraints imposed by the unstable poles of the plant. However, if the YJBK parametrization is used instead, i.e. if $Q(s)$ is the Youla parameter, then even in this case the closed-loop system is stable iff $Q(s)$ is stable.

Remark 3.2.2. Since for an open-loop stable plant, the stability of the configurations of Figs. 3.5 and 3.6 is guaranteed for any stable $Q(s)$, one can now choose $Q(s)$ in an optimal fashion to meet some performance objective. Specific instances of such a choice of $Q(s)$ will be presented in the next section.

3.3 Control Schemes Using the IMC Structure

We now show that different choices of stable $Q(s)$ lead to some familiar control schemes.

3.3.1 Partial Pole Placement Control

From Figure 3.5, it is clear that if the internal model is an exact replica of the plant, then there is no feedback signal in the loop. Consequently the poles of the closed loop system are made up of the open loop poles of the plant and the poles of the IMC parameter $Q(s)$. Thus, in this case, a "complete" pole placement as in traditional pole placement control schemes is not possible. Instead, one can only choose the poles of the IMC parameter $Q(s)$ to be in some desired locations in the left half plane while leaving the remaining closed loop poles at the plant open loop pole locations. Such a control scheme, where $Q(s)$ is chosen to inject an additional set of poles at some desired locations in the complex plane, is referred to as "partial" pole placement.

3.3.2 Model Reference Control

The objective in model reference control is to design a differentiator-free controller so that the output y of the controlled plant $P(s)$ asymptotically tracks the output of a stable reference model $W_m(s)$ for *all* piecewise continuous reference input signals $r(t)$. In order to meet the control objective, we make the following assumptions which are by now standard in the model reference control literature:
(M1) The plant $P(s)$ is minimum phase; and
(M2) The relative degree of the reference model transfer function $W_m(s)$ is greater than or equal to that of the plant transfer function $P(s)$.
Assumption (M1) above is necessary for ensuring internal stability since satisfaction of the model reference control objective requires cancellation of the plant zeros. Assumption (M2) on the other hand permits the design of a differentiator-free controller to meet the control objective. If assumptions (M1) and (M2) are satisfied, it is easy to verify from Figure 3.5 that the choice

$$Q(s) = W_m(s)P^{-1}(s) \tag{3.6}$$

for the IMC parameter guarantees the satisfaction of the model reference control objective in the ideal case, i.e. in the absence of plant modeling errors.

3.3.3 H_2 Optimal Control

In H_2 optimal control, one chooses $Q(s)$ to minimize the L_2 norm of the tracking error $r - y$ provided $r - y \in L_2$. From Figure 3.5, we obtain

$$\begin{aligned} y &= P(s)Q(s)[r] \\ \Rightarrow r - y &= [1 - P(s)Q(s)][r] \\ \Rightarrow \int_0^\infty (r(\tau) - y(\tau))^2 d\tau &= (\|[1 - P(s)Q(s)]R(s)\|_2)^2 \\ &\quad \text{(using Parseval's Theorem)} \end{aligned}$$

where $R(s)$ is the Laplace Transform of $r(t)$ and $\|.(s)\|_2$ denotes the standard H_2 norm. Thus the mathematical problem of interest here is to choose $Q(s)$ to minimize $\|[1 - P(s)Q(s)]R(s)\|_2$. We now derive the analytical expression for the minimizing $Q(s)$. Our derivation closely follows that of [30].

Theorem 3.3.1. *Let $P(s)$ be the stable plant to be controlled and let $R(s)$ be the Laplace Transform of the external input signal $r(t)$*[1]*. Suppose that $R(s)$ has no poles in the* open *right half plane*[2] *and that there exists at least one choice, say $Q_0(s)$, of the stable IMC parameter $Q(s)$ such that $[1 - P(s)Q_0(s)]R(s)$ is stable*[3]*. Let $z_{p_1}, z_{p_2}, \cdots, z_{p_l}$ be the open right half plane zeros of $P(s)$ and define the Blashcke product*[4]

$$B_P(s) = \frac{(-s + z_{p_1})(-s + z_{p_2}) \cdots (-s + z_{p_l})}{(s + \bar{z}_{p_1})(s + \bar{z}_{p_2}) \cdots (s + \bar{z}_{p_l})}$$

so that $P(s)$ can be rewritten as

$$P(s) = B_P(s)P_M(s)$$

where $P_M(s)$ is minimum phase. Similarly, let $z_{r_1}, z_{r_2}, \cdots, z_{r_k}$ be the open right half plane zeros of $R(s)$ and define the Blashcke product

$$B_R(s) = \frac{(-s + z_{r_1})(-s + z_{r_2}) \cdots (-s + z_{r_k})}{(s + \bar{z}_{r_1})(s + \bar{z}_{r_2}) \cdots (s + \bar{z}_{r_k})}$$

so that $R(s)$ can be rewritten as

$$R(s) = B_R(s)R_M(s)$$

---

[1] For the sake of simplicity, both $P(s)$ and $R(s)$ are assumed to be rational transfer functions. The theorem statement can be appropriately modified for the case where $P(s)$ and/or $R(s)$ contain all-pass time delay factors [30].

[2] This assumption is reasonable since otherwise the external input would be unbounded.

[3] The final construction of the H_2 optimal controller serves as proof for the existence of a $Q_0(s)$ with such properties.

[4] Here $\bar{(.)}$ denotes complex conjugation.

where $R_M(s)$ is minimum phase. Then the $Q(s)$ which minimizes $\|[1 - P(s)Q(s)]R(s)\|_2$ is given by

$$Q(s) = P_M^{-1}(s)R_M^{-1}(s)\left[B_P^{-1}(s)R_M(s)\right]_* \tag{3.7}$$

where $[.]_*$ denotes that after a partial fraction expansion, the terms corresponding to the poles of $B_P^{-1}(s)$ are removed.

Proof. Now $\|[1 - P(s)Q(s)]R(s)\|_2$

$$\begin{aligned}
&= \|[1 - P(s)Q_0(s)]R(s) - P(s)[Q(s) - Q_0(s)]R(s)\|_2 \\
&= \|[1 - P(s)Q_0(s)]B_R(s)R_M(s) \\
&\quad - B_P(s)P_M(s)[Q(s) - Q_0(s)]B_R(s)R_M(s)\|_2 \\
&= \frac{1}{2\pi}\int_{-\infty}^{\infty} |[1 - P(j\omega)Q_0(j\omega)] - B_P(j\omega)P_M(j\omega)[Q(j\omega) - Q_0(j\omega)]|^2 \\
&\quad |B_R(j\omega)R_M(j\omega)|^2\, d\omega \\
&= \frac{1}{2\pi}\int_{-\infty}^{\infty} |[1 - P(j\omega)Q_0(j\omega)]B_P^{-1}(j\omega)R_M(j\omega) \\
&\quad - P_M(j\omega)[Q(j\omega) - Q_0(j\omega)]R_M(j\omega)|^2\, d\omega \\
&\quad (\text{since } |B_R(j\omega)| = |B_P(j\omega)| = 1\ \forall\ \omega) \\
&= \|[1 - P(s)Q_0(s)]R_M(s)B_P^{-1}(s) - P_M(s)[Q(s) - Q_0(s)]R_M(s)\|_2
\end{aligned} \tag{3.8}$$

Since $P_M(s)$, $Q(s)$, $Q_0(s)$ are all stable, and the only unstable poles of $R_M(s)$ must lie on the imaginary axis, it follows that the right half plane poles of $P_M(s)[Q(s) - Q_0(s)]R_M(s)$, if any, must lie on the imaginary axis. On the other hand, the assumption about $[1 - P(s)Q_0(s)]R(s)$ being stable implies that $[1 - P(s)Q_0(s)]R_M(s)$ is stable which in turn implies that $[1 - P(s)Q_0(s)]R_M(s)B_P^{-1}(s)$ can have unstable poles only in the open right half plane. Hence, using the fact that the frequency domain L_2 norm[5] is well-defined only for transfer functions having no imaginary axis poles, it follows that $P_M(s)[Q(s) - Q_0(s)]R_M(s)$ is stable. This implies that the expression in (3.8) is minimized when

$$P_M(s)[Q(s) - Q_0(s)]R_M(s) = \left\{B_P^{-1}(s)[1 - P(s)Q_0(s)]R_M(s)\right\}_{\infty+}$$

where $\{.\}_{\infty+}$ denotes that after a partial fraction expansion, only the constant term and the terms corresponding to the open left half plane poles are retained. The above result is obtained as a consequence of the Projection Theorem in Hilbert spaces [27]. The optimal $Q(s)$ is, therefore, given by

$$Q(s) = Q_0(s) + P_M^{-1}(s)R_M^{-1}(s)\left\{B_P^{-1}(s)[1 - P(s)Q_0(s)]R_M(s)\right\}_{\infty+}. \tag{3.9}$$

[5] The freqency domain L_2 norm of a strictly proper transfer function $H(s)$ with no poles on the imaginary axis is given by $\frac{1}{\sqrt{2\pi}}\left\{\int_{-\infty}^{\infty} |H(j\omega)|^2 d\omega\right\}^{\frac{1}{2}}$.

We now carry out some algebra to simplify the expression for $Q(s)$ and render it independent of the particular choice of $Q_0(s)$. Equation (3.9) can be rewritten as

$$\begin{aligned} Q(s) &= P_M^{-1}(s)R_M^{-1}(s)\Big[P_M(s)R_M(s)Q_0(s) + \left\{B_P^{-1}(s)R_M(s)\right\}_{\infty+} \\ &\quad - \left\{B_P^{-1}(s)P(s)Q_0(s)R_M(s)\right\}_{\infty+}\Big] \\ &= P_M^{-1}(s)R_M^{-1}(s)\left[\{P_M(s)R_M(s)Q_0(s)\}_{0-} \right. \\ &\quad \left. + \left\{B_P^{-1}(s)R_M(s)\right\}_{\infty+}\right] \end{aligned}$$

where $\{.\}_{0-}$ denotes that after a partial fraction expansion, only the terms having poles in the closed right half plane are retained.

$$\begin{aligned} &= P_M^{-1}(s)R_M^{-1}(s)\Big[\left\{P(s)Q_0(s)B_P^{-1}(s)R_M(s)\right\}_{0-} \\ &\quad + \left\{B_P^{-1}(s)R_M(s)\right\}_{\infty+}\Big] \\ &= P_M^{-1}(s)R_M^{-1}(s)[B_P^{-1}(s)R_M(s)]_* \end{aligned}$$

(since the stability requirement on $(1 - P(s)Q_0(s))R(s)$ mandates that $P(s)Q_0(s) = 1$ at the imaginary axis poles of $R_M(s)$)

This completes the proof.

Remark 3.3.1. The optimal $Q(s)$ defined in (3.7) is usually improper. So it is customary to make $Q(s)$ proper by introducing sufficient high frequency attenuation via what is called the "IMC Filter" $F(s)$ [30]. The design of the IMC Filter depends on the choice of the input $R(s)$. Although this design is carried out in a somewhat adhoc fashion, care is taken to ensure that the original asymptotic tracking properties of the controller are preserved. Instead of the optimal $Q(s)$ in (3.7), the $Q(s)$ to be implemented is given by

$$Q(s) = P_M^{-1}(s)R_M^{-1}(s)\left[B_P^{-1}(s)R_M(s)\right]_* F(s) \tag{3.10}$$

where $F(s)$ is the stable IMC Filter. As a specific example, suppose that the system is of Type 1[6]. Then, a possible choice for the IMC Filter is $F(s) = \frac{1}{(\tau s+1)^{n^*}}$, $\tau > 0$ where n^* is chosen to be a large enough positive integer to make $Q(s)$ proper. As shown in [30], the parameter τ represents a tradeoff between tracking performance and robustness to modelling errors.

3.3.4 H_∞ Optimal Control

The sensitivity function $S(s)$ and the complementary sensitivity function $T(s)$ for the IMC configuration in Figure 3.5 are given by $S(s) = 1 - P(s)Q(s)$

[6] Other system types can also be handled as in [30].

and $T(s) = P(s)Q(s)$ respectively [30] (also see Chapter 1, Section 1.3). Since the plant $P(s)$ is open loop stable, it follows that the H_∞ norm of the complementary sensitivity function $T(s)$ can be made *arbitrarily small* by simply choosing $Q(s) = \frac{1}{k}$ and letting k tend to infinity. Thus minimizing the H_∞ norm of $T(s)$ does not make much sense since the infimum value of zero is unattainable.

On the other hand, if we consider the weighted sensitivity minimization problem where we seek to minimize $||W(s)S(s)||_\infty$ for some stable, rational, minimum phase[7], weighting transfer function $W(s)$, then we have an interesting H_∞ minimization problem i.e. choose a stable $Q(s)$ to minimize $||W(s)[1-P(s)Q(s)]||_\infty$. The solution to this problem depends on the number of open right half plane zeros of the plant $P(s)$. For instance, when the plant has no right half-plane zeros, then by choosing $Q(s) = P^{-1}(s)/((\tau s+1)^{n^*})$ where n^* is the relative degree of the plant and by letting $\tau > 0$ tend to zero, we can make $||W(s)[1-P(s)Q(s)]||_\infty$ arbitrarily small [47]. Thus in this case also we have an unattainable infimum of zero and the H_∞ norm minimization problem is not interesting. However, when the plant $P(s)$ has one or more right half-plane zeros, say b_i, then the stability constraint on $Q(s)$ mandates that $P(b_i)Q(b_i) = 0$ which implies that $W(b_i)[1 - P(b_i)Q(b_i)] = W(b_i)$. This is referred to as an "interpolation constraint" for the sensitivity function and, as will be seen shortly, it imposes a lower bound on the attainable magnitude of the weighted sensitivity function.

The general weighted sensitivity minimization problem involving multiple right half-plane zeros is quite mathematically involved and requires the use of Nevanlinna Pick interpolation [48]. However, when the plant has only one right half-plane zero b_1 and none on the imaginary axis, there is only one interpolation constraint and a closed form solution can be obtained by using the following property derived from the *Maximum Modulus Theorem* of complex analysis [38].

Property 3.3.1. For any transfer function which is proper and *stable*, the magnitude of the value of the transfer function evaluated anywhere in the open right half-plane is bounded from above by the maximum magnitude on the imaginary axis.

Since $W(s)[1 - P(s)Q(s)]$ is stable, it follows using the above property that

$$||W(s)[1 - P(s)Q(s)]||_\infty \geq |W(b_1)|. \tag{3.11}$$

Choosing

$$Q(s) = [1 - \frac{W(b_1)}{W(s)}]P^{-1}(s). \tag{3.12}$$

[7] The minimum phase assumption is without any loss of generality since given any non-minimum phase stable weighting transfer function, one can always find a minimum phase weighting $W(s)$ whose magnitude on the imaginary axis, and hence the H_∞ norm weighting, is identical to that produced by the original weighting transfer function.

we see that equality is attained in (3.11) so that (3.12) is indeed the expression for the optimal $Q(s)$.

Fortunately, the case just considered covers a large number of process control applications where plants are typically modelled as minimum phase first or second order transfer functions with time delays. Since approximating a delay using a first order Pade approximation introduces one right half plane zero, the resulting rational approximation will satisfy the one right half plane zero assumption.

Remark 3.3.2. As in the case of H_2 optimal control, the optimal $Q(s)$ defined by (3.12) is usually improper. This situation can be handled as in Remark 3.3.1 so that the $Q(s)$ to be implemented becomes

$$Q(s) = [1 - \frac{W(b_1)}{W(s)}]P^{-1}(s)F(s) \tag{3.13}$$

where $F(s)$ is a stable IMC Filter as before.

3.3.5 Robustness to Modelling Errors

In a later chapter, we will be combining the above IMC schemes with a robust adaptive law to obtain robust adaptive IMC schemes. If the above IMC schemes, which are based on the assumption of known plant parameters, are themselves unable to tolerate uncertainty, then there is little or no hope that adaptive designs based on them will do any better when additionally the plant parameters are unknown and have to be estimated using an adaptive law. Accordingly, we now examine the robustness of the non-adaptive IMC schemes discussed above to the presence of plant modelling errors. Without any loss of generality let us suppose that the uncertainty is of the multiplicative type i.e.

$$P(s) = P_0(s)(1 + \mu\Delta_m(s)) \tag{3.14}$$

where $P_0(s)$ is the modelled part of the plant and $\mu\Delta_m(s)$ is a stable multiplicative uncertainty such that $P_0(s)\Delta_m(s)$ is strictly proper. Then we can state the following robustness result which follows immediately from the Small Gain Theorem (see Example 2.3.1). An alternative proof can also be found in [30].

Theorem 3.3.2. *Suppose $P_0(s)$ and $Q(s)$ are stable transfer functions so that the IMC configuration in Figure 3.5 is stable for $P(s) = P_0(s)$. Then the IMC configuration with the actual plant given by (3.14) is still stable provided $\mu \in [0, \mu^*)$ where $\mu^* = \frac{1}{\|P_0(s)Q(s)\Delta_m(s)\|_\infty}$.*

Remark 3.3.3. Theorem 3.3.2 shows that the non-adaptive IMC schemes considered in this chapter are robust with respect to modelling errors in the plant. This gives us good reason to believe that adaptive versions of the above schemes could also be designed to be robust to modelling errors, provided the adaptive law or parameter estimator is chosen appropriately [19]. A step-by-step development leading to such a design will be presented in the chapters to follow.

CHAPTER 4
ON-LINE PARAMETER ESTIMATION

4.1 Introduction

In this chapter, we focus on the design of on-line parameter estimators or adaptive laws. An adaptive control scheme has two essential ingredients: (i) a control structure that could be used to meet the control objective if the plant parameters are known; and (ii) a parameter estimator or adaptive law which provides estimates of the plant or controller parameters, to be used when the plant parameters and hence the controller parameters are unknown. The properties of both the control structure and the adaptive law play a crucial role in determining the stability of the closed loop adaptive system. Having already studied some of the control structures of interest to us in the last chapter, we now concentrate on the adaptive laws.

For clarity of presentation, in this chapter, we will make the unrealistic assumption that the plant to be estimated is a perfectly modelled one, i.e. there are no modelling errors. This will allow us to gain a thorough understanding of the principles involved in designing adaptive laws, without getting mired in the mathematical intricacies that are required for "robustifying" these algorithms. Of courses, these issues are very important from a practical point of view and will be taken up in a later chapter.

The design of adaptive laws in this chapter is introduced via some simple examples. The approach then extends to the case of a general plant by making use of a certain *linear parametric model.* The need to design adaptive laws which can guarantee certain properties, regardless of the control objective leads to the introduction of what is called "normalization" into the adaptive law. Again, the normalization procedure and its associated analysis are first illustrated on the simple examples before being applied to the general plant. Finally, the chapter concludes with a detailed treatment of how to incorporate *apriori knowledge* of the whereabouts of the unknown parameters into the adaptive law. This is done via the mechanism of *parameter projection.*

4.2 Simple Examples

In this section, we use simple examples to illustrate the procedure involved in designing and analyzing on-line parameter estimators. We first consider a scalar example with one unknown parameter and no dynamics and then a first order example with two unknown parameters.

4.2.1 Scalar Example: One Unknown Parameter

Consider the simplest possible parameter estimation problem: obtain an estimate of the unknown scalar parameter θ^* from the measurements of $y(t)$ and $u(t)$ where $y : R^+ \mapsto R$, $u : R^+ \mapsto R$ are related by

$$y(t) = \theta^* u(t) \; \forall \, t \geq 0 \tag{4.1}$$

A rather naive approach to solving this problem would be to obtain the estimate $\theta(t)$ of θ^* at time t by simply calculating the quotient $y(t)/u(t)$. However, such an approach is not acceptable because of at least two reasons. First, the measurements of $y(t)$ and $u(t)$ will invariably be corrupted with noise making the estimate $\theta(t)$ very inaccurate; and second, the signal $u(t)$ may pass through zero for certain values of t for which the estimate $\theta(t)$ as calculated above will be undefined.

An alternative approach to the problem is to proceed as follows. Let $\theta(t)$ be an estimate of θ^* at time t. Then since $u(t)$ is available for measurement, in view of (4.1), we can construct an estimate $\hat{y}(t)$ of $y(t)$ as

$$\hat{y}(t) = \theta(t)u(t) \tag{4.2}$$

Now construct the signal

$$\epsilon_1(t) = y(t) - \hat{y}(t) = y(t) - \theta(t)u(t) \tag{4.3}$$

and observe that if $\theta(t) = \theta^*$, i.e. our estimate is in fact equal to the true value, then $\epsilon_1(t) = 0$. Thus the signal ϵ_1 is a measure of the quality of our estimate and is, therefore, referred to as the *estimation error*. Many parameter estimators can now be derived by considering various cost functions involving the estimation error and using iterative techniques for minimizing them. One such parameter estimator can be derived by considering the instantaneous cost function

$$J(t, \theta(t)) = \epsilon_1^2(t)/2 = ((y(t) - \theta(t)u(t))^2)/2 \tag{4.4}$$

and minimizing J with respect to θ using the *Gradient Method*.

For such a minimization problem to be well posed, it is necessary that the cost function J be such that it cannot be driven to infinity by any variable other than θ. This can be guaranteed by assuming that $u \in L_\infty$, an assumption that will be made throughout this subsection[1].

---

[1] As an aside, we note that, under this assumption, we also have that for each time t, $J(t, \theta(t))$ is a convex function over R^1. This fact can be used to minimize J with respect to θ using *Newton's Method* and results in a least squares parameter estimator [19].

A brief intuitive treatment of the gradient method is included in Appendix B. From (B.4) in Appendix B, it follows that the parameter estimator for this case is given by

$$\dot{\theta} = -\gamma \nabla J(\theta), \ \theta(0) = \theta_0, \ \gamma > 0 \tag{4.5}$$

where γ is a design parameter called the *adaptive gain*. However, from (4.4), $\nabla J(\theta) = -\epsilon_1 u$ so that (4.5) becomes

$$\dot{\theta} = \gamma \epsilon_1 u, \ \theta(0) = \theta_0, \ \gamma > 0 \tag{4.6}$$

Equation (4.6) above is the *standard gradient parameter estimator* for estimating θ^* in the model (4.1). We next proceed to analyze its properties.

Defining $\tilde{\theta} \triangleq \theta - \theta^*$, and noting that θ^* is a constant, the adaptive law (4.6) can be written as

$$\dot{\tilde{\theta}} = \gamma \epsilon_1 u, \ \tilde{\theta}(0) = \theta(0) - \theta^* \tag{4.7}$$

Furthermore, substituting (4.1) in (4.3), we obtain

$$\epsilon_1 = \theta^* u - \theta u = -\tilde{\theta} u \tag{4.8}$$

Substituting (4.8) into (4.7), it follows that

$$\dot{\tilde{\theta}} = -\gamma u^2 \tilde{\theta} \tag{4.9}$$

which is a differential equation with equilibrium state $\tilde{\theta}_e = 0$. Now consider the positive definite function

$$V(\tilde{\theta}) = \tilde{\theta}^2/(2\gamma)$$

whose derivative along the solution of (4.7) is given by

$$\begin{aligned} \dot{V} = \tilde{\theta}(\epsilon_1 u) &= -\epsilon_1^2 \text{ (using (4.8))} \\ &\leq 0. \end{aligned} \tag{4.10}$$

Hence, it follows from Theorem 2.4.1 that the equilibrium $\tilde{\theta}_e = 0$ of (4.9) is uniformly stable. Thus $\tilde{\theta}, \theta \in L_\infty$ and from (4.8), (4.7) $\epsilon_1, \dot{\theta} \in L_\infty$. Furthermore, V is bounded from below (by zero) and, in view of (4.10), is monotonically non-increasing. Hence, by Lemma 2.2.5(i), $V_\infty \triangleq \lim_{t\to\infty} V(\tilde{\theta}(t))$ exists. Therefore, integrating both sides of (4.10) from 0 to t and letting t tend to infinity, we conclude that $\epsilon_1 \in L_2$, which in view of (4.7) implies that $\dot{\theta} \in L_2$. If, in addition, $\dot{u} \in L_\infty$, then $\dot{\epsilon}_1 = -\dot{\tilde{\theta}} u - \tilde{\theta}\dot{u} = -\gamma \epsilon_1 u^2 - \tilde{\theta}\dot{u} \in L_\infty$ and using Corollary 2.2.1, it follows that $\epsilon_1(t) \to 0$ as $t \to \infty$. This, in turn, implies that $\dot{\theta} = \gamma \epsilon_1 u \to 0$ as $t \to \infty$.

Note that the above analysis was carried out under the assumption that $u, \dot{u} \in L_\infty$. No further assumptions were made on the input signal u. An important question that now arises is whether $\theta(t)$ converges to θ^* as $t \to \infty$.

The answer depends very much on the properties of the input signal u. The following analysis clearly brings this out.

Now, from (4.9), we obtain

$$\tilde{\theta}(t) = e^{-\gamma \int_0^t u^2(\tau)d\tau} \tilde{\theta}(0).$$

Clearly, if $u(t) = 0$ or $u(t) = e^{-t}$, then $\tilde{\theta}(t)$ does not tend to zero as $t \to \infty$, i.e. parameter convergence does not take place. If, on the other hand, $u(t) = 1/(\sqrt{1+t})$, then $\tilde{\theta}(t)$ tends to zero asymptotically but not exponentially. Finally, if $u(t) =$ a non-zero constant, 1 say, then $\tilde{\theta}(t)$ goes to zero exponentially fast. It can be shown that a necessary and sufficient condition for $\tilde{\theta}(t)$ in this example to converge to zero exponentially fast is that $u(t)$ satisfies

$$\int_t^{t+T_0} u^2(\tau)d\tau \geq \alpha_0 T_0$$

$\forall\ t \geq 0$ and for some $\alpha_0, T_0 > 0$. This property of u is referred to as *persistence of excitation* (PE) and u is said to be *persistently exciting*. The positive constant α_0 is called the level of excitation.

Remark 4.2.1. The study of persistent excitation is very important in parameter identification problems where parameter convergence is crucial. However, in this monograph, we are interested in designing on-line parameter estimators for use in adaptive *control* schemes, where parameter convergence is *not* part of the control objective. Accordingly, we will not discuss parameter convergence in any detail, especially since such a discussion would be very technical and mathematically involved. The interested reader is referred to Chapter 4, Section 8 of [19] for an exhaustive treatment.

Remark 4.2.2. In this subsection, we have designed and analyzed the standard gradient parameter estimator based on the instantaneous cost. As already indicated, there is nothing unique about this cost function or the minimization method used. Indeed, as shown in [15], [19], it is possible to derive other adaptive laws by changing the cost function and/or using other methods of minimization such as the Newton's Method. However, for our purposes, we will stick to the gradient method applied to the instantaneous cost. This is consistent with our objective which is to only familiarize industrial practitioners with the methodology for designing adaptive internal model control schemes. Once such familiarization is achieved, we believe that the reader should be able to look at the existing adaptive literature e.g. [19] and easily replace the gradient update law by other update laws such as least squares.

The parameter estimation problem considered in this subsection involves only one scalar unknown θ^* and a static, i.e. algebraic relationship between the measured signals $u(t), y(t)$. In the next subsection, we will consider the parameter estimation problem where there are two unknown parameters and the measured signals are dynamically related.

4.2.2 First Order Example: Two Unknowns

Consider the first order plant

$$\dot{y} = -ay + bu, \ y(0) = y_0 \tag{4.11}$$

where the parameters a and b are unknown, $a > 0$, and the input $u(t)$ and the output $y(t)$ are signals that are available for measurement. The estimation problem here is to determine on-line estimates of the parameters a and b from the measurements of $y(t)$ and $u(t)$. *If the signal $\dot{y}(t)$ were available for measurement*, then one could define $z = \dot{y}$, $\theta^* = [-a, b]^T$, $\phi = [y, u]^T$ so that (4.11) could be rewritten as

$$z = {\theta^*}^T \phi \tag{4.12}$$

where the signal ϕ is called the *regressor vector*. Equation (4.12) is exactly in the same form as (4.1) with z and ϕ taking the place of y and u. Consequently, one could proceed as in the previous subsection to design an on-line parameter estimator for θ^*. Specifically, let $\theta(t)$ be the estimate of θ^* at time t. Then, in view of (4.12), one can use this estimate to construct the estimate $\hat{z}$ of z as

$$\hat{z} = \theta^T \phi. \tag{4.13}$$

Define the estimation error $\epsilon_1 = z - \hat{z} = z - \theta^T \phi$ and, as before, consider the cost function $J = \epsilon_1^2/2 = (z - \theta^T \phi)^2/2$. If we now make the assumption that ϕ is bounded, then J cannot be driven to infinity by z or by ϕ. Hence, minimizing J with respect to θ is a meaningful optimization problem. Now, in this case, $\nabla J(\theta) = -\epsilon_1 \phi$, so from the gradient method we obtain

$$\dot{\theta} = \gamma \epsilon_1 \phi, \ \gamma > 0, \ \theta(0) = \theta_0 \tag{4.14}$$

as the gradient parameter estimator for estimating a and b. Since $a > 0$, it follows that (4.11) represents a stable plant so that the boundedness of ϕ is guaranteed as long as $u(t)$ is chosen to be bounded. Using a similar procedure as in the last subsection, one could now go ahead and analyze the properties of the adaptive law (4.14). However, the adaptive law (4.14) is not implementable since $\dot{y}$ is not available for measurement and any attempt to differentiate the measured signal $y(t)$ to obtain $\dot{y}$ will lead to undesirable amplification of the high frequency noise, leading to a poor signal-to-noise ratio (SNR).

This problem can be avoided altogether by filtering all the signals in (4.11) before attempting to design the adaptive law. For instance, by filtering both sides of (4.11) using the stable filter $\Lambda(s) = 1/(s+1)$ we obtain

$$\frac{s}{s+1}[y] = -a\frac{1}{s+1}[y] + b\frac{1}{s+1}[u] \tag{4.15}$$

Defining $z = \frac{s}{s+1}[y]$, $\phi = \left[\frac{1}{s+1}[y], \frac{1}{s+1}[u]\right]^T$ and $\theta^* = [-a, b]^T$, we can rewrite (4.15) as

$$z = \theta^{*T}\phi \tag{4.16}$$

Since the signals z and ϕ in (4.16) are available for measurement and ϕ can be guaranteed to be bounded by assuming that u is bounded, we can start from (4.16) and proceed as before to obtain the gradient parameter estimator

$$\dot{\theta} = \gamma\epsilon_1\phi,\ \gamma > 0,\ \theta(0) = \theta_0 \tag{4.17}$$

$$\text{where } \epsilon_1 = z - \theta^T\phi. \tag{4.18}$$

We next analyze the properties of this adaptive law. The analysis is in the same spirit as the corresponding one in the last subsection. Define $\tilde{\theta} \triangleq \theta - \theta^*$ and consider the Lyapunov-like function $V(\tilde{\theta}) = \frac{\tilde{\theta}^T\tilde{\theta}}{2\gamma}$. Then taking the time derivative of V along the solution of (4.17), we obtain

$$\begin{aligned}\dot{V} &= \epsilon_1\tilde{\theta}^T\phi \\ &= -\epsilon_1^2 \text{ (using (4.16) and (4.18))} \\ &\leq 0.\end{aligned}$$

Hence, it follows that $\tilde{\theta}, \theta$ and ϵ_1 are bounded. Since V is monotonically decreasing and bounded from below, $V_\infty = \lim_{t\to\infty} V(\tilde{\theta}(t))$ exists. So, integrating both sides of $\dot{V} = -\epsilon_1^2$ from 0 to ∞, we conclude that $\epsilon_1 \in L_2$. Now, if $\dot{\phi}$ is bounded then

$$\dot{\epsilon}_1 = \frac{d}{dt}(-\tilde{\theta}^T\phi) = -\dot{\tilde{\theta}}^T\phi - \tilde{\theta}^T\dot{\phi} = -\gamma\epsilon_1\phi^T\phi - \tilde{\theta}^T\dot{\phi}$$

is bounded and, by Corollary 2.2.1, it follows that $\epsilon_1(t) \to 0$ as $t \to \infty$. This in turn implies that that $\dot{\theta} = \gamma\epsilon_1\phi \to 0$ as $t \to \infty$. This completes the analysis of the properties of (4.17). As in the last subsection, this analysis does not tell us anything about the convergence, if any, of the parameter estimates.

Remark 4.2.3. The design of the adaptive law (4.17) was based on rewriting the filtered plant (4.15) as (4.16). Such a representation is by no means unique. For instance, one could alternatively define $z = y$, $\phi = \left[\frac{1}{s+1}[y], \frac{1}{s+1}[u]\right]^T$ and $\theta^* = [-a+1, b]^T$ and obtain the same kind of representation as in (4.16). The design and analysis of the adaptive law for estimating the new θ^* would also proceed along exactly the same steps. As we will see in the next chapter, this particular plant representation fits in very naturally with the IMC structure and is, therefore, very useful for designing adaptive IMC schemes. Consequently, this representation will be discussed in detail in the next section when considering the problem of estimating the parameters of a general plant.

4.3 The General Case

Consider the general linear time-invariant finite-dimensional plant described by the input-output differential equation

$$y^{(n)} + a_1 y^{(n-1)} + a_2 y^{(n-2)} + \cdots + a_n y = b_0 u^{(l)} + b_1 u^{(l-1)} + \cdots + b_l u \quad (4.19)$$

where $n > l$ and $u(t)$, $y(t)$ represent the input and output signals of the plant respectively. Taking Laplace transforms on both sides of (4.19) and setting the initial conditions to zero, the plant (4.19) can also be represented as

$$y = \frac{Z_0(s)}{R_0(s)}[u] \quad (4.20)$$

where $\frac{Z_0(s)}{R_0(s)}$ is the transfer function of the plant and $R_0(s) = s^n + a_1 s^{n-1} + \cdots + a_{n-1}s + a_n$, $Z_0(s) = b_0 s^l + b_1 s^{l-1} + \cdots + b_l$. The parameter estimation problem here is to obtain estimates of the unknown parameters $a_1, a_2, \cdots, a_n, b_0, b_1, \cdots, b_l$ based on measurements of only the plant input u and the plant output y. Since the derivatives of the plant input and output are not available for measurement, we filter both sides of (4.19) with $\frac{1}{\Lambda(s)}$, where $\Lambda(s)$ is an arbitrary monic Hurwitz polynomial of degree n, to obtain

$$\begin{aligned} &\frac{s^n}{\Lambda(s)}[y] + a_1 \frac{s^{n-1}}{\Lambda(s)}[y] + a_2 \frac{s^{n-2}}{\Lambda(s)}[y] + \cdots + a_n \frac{1}{\Lambda(s)}[y] = \\ &b_0 \frac{s^l}{\Lambda(s)}[u] + b_1 \frac{s^{l-1}}{\Lambda(s)}[u] + \cdots + b_l \frac{1}{\Lambda(s)}[u] \end{aligned} \quad (4.21)$$

Defining

$$z = \frac{s^n}{\Lambda(s)}[y] \quad (4.22)$$

$$\theta^* = [-a_1, -a_2, \cdots, -a_n, b_0, b_1, \cdots, b_l]^T \quad (4.23)$$

$$\phi = \left[\frac{s^{n-1}}{\Lambda(s)}[y], \frac{s^{n-2}}{\Lambda(s)}[y], \cdots, \frac{1}{\Lambda(s)}[y], \frac{s^l}{\Lambda(s)}[u], \frac{s^{l-1}}{\Lambda(s)}[u], \cdots, \frac{1}{\Lambda(s)}[u]\right]^T \quad (4.24)$$

we can represent (4.21) in the form

$$z = \theta^{*T} \phi \quad (4.25)$$

Now, if ϕ is bounded, then we can proceed as in the last subsection to design and analyze a standard gradient parameter estimator. The boundedness of ϕ can be guaranteed provided we choose the input u to be bounded and the plant transfer function $\frac{Z_0(s)}{R_0(s)}$ is assumed to be stable.

Now let us consider a slightly different representation of the filtered plant (4.21). Adding y to both sides of (4.21) and re-arranging terms, we obtain

$$\begin{aligned} y &= \frac{\Lambda(s)}{\Lambda(s)}[y] - \frac{s^n}{\Lambda(s)}[y] - a_1\frac{s^{n-1}}{\Lambda(s)}[y] - a_2\frac{s^{n-2}}{\Lambda(s)}[y] + \cdots - a_n\frac{1}{\Lambda(s)}[y] + \\ & \quad b_0\frac{s^l}{\Lambda(s)}[u] + b_1\frac{s^{l-1}}{\Lambda(s)}[u] + \cdots + b_l\frac{1}{\Lambda(s)}[u] \end{aligned} \tag{4.26}$$

Let $\Lambda(s) = s^n + \lambda_1 s^{n-1} + \lambda_2 s^{n-2} + \cdots + \lambda_n$. Then defining

$$z = y \tag{4.27}$$

$$\theta^* = [\lambda_1 - a_1, \lambda_2 - a_2, \cdots, \lambda_n - a_n, b_0, b_1, \cdots, b_l]^T \tag{4.28}$$

and with ϕ defined as in (4.24), we once again obtain the plant representation

$$z = \theta^{*T}\phi \tag{4.29}$$

Remark 4.3.1. The plant representations (4.25) or (4.29) are referred to as *linear parametric models* because the unknown parameter vector θ^* appears linearly in each one of them. Such models are very popular in the parameter estimation literature since adaptive laws based on them *can be proven* to guarantee certain properties. Although the design and analysis of adaptive laws based on (4.25) or (4.29) is exactly the same with only z and θ^* being defined differently, in this monograph, we will specifically focus on the model (4.29) *since it interfaces in a natural fashion with the IMC structure.* Furthermore, as indicated earlier, we will concentrate only on adaptive laws designed using the gradient method. The interested reader is referred to [19] for an exhaustive treatment of the other kinds of models that are used in parameter estimation and the different approaches that are used for deriving the adaptive laws.

We now design and analyze the gradient parameter estimator based on the model (4.29). The steps are the same as in the design and analysis for the first order example presented in the last section. However, we repeat the detailed derivation here since the general case being now considered is the one that will be referred to in the later chapters. Let $\theta(t)$ be the estimate of θ^* at time t. Then in view of (4.29), the estimate $\hat{z}$ of z is constructed as

$$\hat{z} = \theta^T\phi \tag{4.30}$$

Define the estimation error

$$\epsilon_1 = z - \hat{z} = z - \theta^T\phi \tag{4.31}$$

and consider the instantaneous cost function $J = \frac{\epsilon_1^2}{2} = \frac{(z-\theta^T\phi)^2}{2}$. Now assume that the plant $\frac{Z_0(s)}{R_0(s)}$ is stable and the input u is chosen to be bounded. Then the signals z and ϕ in (4.29) are certainly bounded so that the cost function

J cannot be driven to infinity by z or by ϕ. Hence, minimizing J with respect to θ is a well posed optimization problem. Now, from the expression for J, $\nabla J(\theta) = -\epsilon_1 \phi$. Hence, using the Gradient Method, we obtain

$$\dot{\theta} = \gamma \epsilon_1 \phi, \ \gamma > 0, \ \theta(0) = \theta_0 \tag{4.32}$$

as the gradient update law.

The following Theorem summarizes the properties of the adaptive law (4.32).

Theorem 4.3.1. *Consider the gradient update law (4.32) based on the linear parametric model (4.29) where the signal* ϕ *is known to be bounded. Then* $\theta, \epsilon_1 \in L_\infty$ *and* $\epsilon_1 \in L_2$. *If, in addition,* $\dot{\phi} \in L_\infty$, *then* $\epsilon_1(t) \to 0$ *as* $t \to \infty$.

Proof. Define the parameter error $\tilde{\theta} = \theta - \theta^*$ and consider the Lyapunov-like function $V(\tilde{\theta}) = \frac{\tilde{\theta}^T \tilde{\theta}}{2\gamma}$. Then along the solution of (4.32), we have

$$\begin{aligned} \dot{V} &= \tilde{\theta}^T \epsilon_1 \phi \\ &= -\epsilon_1^2 \text{ (using (4.29) and (4.31))} \\ &\leq 0 \end{aligned} \tag{4.33}$$

Hence it follows that $\tilde{\theta} \in L_\infty$ which in turn implies that $\theta \in L_\infty$. From (4.29) and (4.31), we obtain $\epsilon_1 = -\tilde{\theta}^T \phi$. Since $\tilde{\theta}, \phi \in L_\infty$, it follows that $\epsilon_1 \in L_\infty$.

Now from (4.33), we see that V is a monotonically non-increasing function which is bounded from below. Hence $V_\infty \triangleq \lim_{t\to\infty} V(\tilde{\theta}(t))$ exists and integrating the equation $\dot{V} = -\epsilon_1^2$ on both sides from 0 to ∞, we obtain

$$\int_0^\infty \epsilon_1^2(\tau) d\tau = V(\tilde{\theta}(0)) - V_\infty.$$

Thus $\epsilon_1 \in L_2$. Now, in addition, if $\dot{\phi} \in L_\infty$, then

$$\dot{\epsilon}_1 = \frac{d}{dt}(-\tilde{\theta}^T \phi) = -\dot{\tilde{\theta}}^T \phi - \tilde{\theta}^T \dot{\phi} = -\gamma \epsilon_1 \phi^T \phi - \tilde{\theta}^T \dot{\phi} \in L_\infty.$$

Hence, by Corollary 2.2.1, it follows that $\epsilon_1(t) \to 0$ as $t \to \infty$ and the proof is complete.

4.4 Adaptive Laws with Normalization

The adaptive laws developed so far require that the input u and the output y of the plant be apriori assumed to be bounded. Such an assumption is reasonable for the parameter estimation of stable plants where one is free to choose the input u and, by choosing any bounded u, the satisfaction of these properties is guaranteed. The situation, however, is quite different in adaptive control where neither the input nor the output signal can be apriori assumed to be bounded. Indeed, guaranteeing the boundedness of these signals in the closed loop is usually part of the control objective. Our objective in this section is to examine how the parameter estimator designs presented earlier can be modified to guarantee certain desirable properties even when the plant input and output signals are not apriori assumed to be bounded. As before, we first consider the scalar and first order examples before moving on to the case of a general plant.

4.4.1 Scalar Example

Consider the problem of estimating the unknown scalar parameter θ^* from measurements of $u(t)$ and $y(t)$ where $u : R^+ \mapsto R$, $y : R^+ \mapsto R$ and are related by

$$y(t) = \theta^* u(t) \tag{4.34}$$

Unlike the parameter estimation problem considered in Section 4.2.1, the input $u(t)$ and the output $y(t)$ are now not necessarily bounded. Nevertheless, let us proceed as in Section 4.2.1 and see exactly where we run into difficulties in trying to design a gradient parameter estimator based on the instantaneous cost. To this end, let $\theta(t)$ be the estimate of θ^* at time t. Then the estimate $\hat{y}$ of y is constructed as $\hat{y} = \theta u$ leading to the estimation error $\epsilon_1 = y - \hat{y} = y - \theta u$. If we now consider the instantaneous cost function $J = \frac{\epsilon_1^2}{2} = \frac{(y-\theta u)^2}{2}$ and try to minimize it with respect to θ, we find that we have an ill-posed minimization problem since u and y are not guaranteed to be bounded and could easily drive J to infinity, regardless of θ. The situation could be remedied by proceeding as follows.

Divide each side of (4.34) by some function $m(t) > 0$ referred to as the *normalizing signal* to obtain

$$\bar{y} = \theta^* \bar{u} \tag{4.35}$$

where $\bar{y} = \frac{y}{m}$ and $\bar{u} = \frac{u}{m}$ are the normalized values of y and u respectively, and $m^2 = 1 + n_s^2$. The signal n_s is chosen so that $\frac{u}{m} \in L_\infty$, e.g. $n_s = u$. Since $\bar{u}, \bar{y} \in L_\infty$, we can start with (4.35) and use the same approach as in Section 4.2.1. We next present the detailed steps involved.

Let $\theta(t)$ be the estimate of θ^* at time t. Then, in view of (4.35), we obtain the estimate $\hat{\bar{y}}$ of $\bar{y}$ as

$$\hat{\bar{y}} = \theta \bar{u} \tag{4.36}$$

Define the estimation error $\bar{\epsilon}_1$ by

$$\bar{\epsilon}_1 = \bar{y} - \hat{\bar{y}} = \bar{y} - \theta \bar{u} \tag{4.37}$$

and consider the instantaneous cost function $J = \frac{\bar{\epsilon}_1^2}{2} = \frac{(\bar{y}-\theta\bar{u})^2}{2} = \frac{(y-\theta u)^2}{2m^2}$. Since $\frac{u}{m}, \frac{y}{m} \in L_\infty$, it follows that J cannot be driven to ∞ by y or u. Hence, minimizing J with respect to θ is a meaningful optimization problem and using the gradient method, we obtain the adaptive law

$$\dot{\theta} = \gamma \bar{\epsilon}_1 \bar{u}, \ \gamma > 0, \ \theta(0) = \theta_0 \tag{4.38}$$

The adaptive law (4.38) can be rewritten in terms of the unnormalized signals ϵ_1 and u by substituting $\bar{\epsilon}_1 = \frac{\epsilon_1}{m}$, $\bar{u} = \frac{u}{m}$. This leads to

$$\dot{\theta} = \frac{\gamma \epsilon_1 u}{m^2}, \ \theta(0) = \theta_0 \tag{4.39}$$

As already mentioned, a possible choice for m which guarantees that $\frac{u}{m} \in L_\infty$ is $m^2 = 1 + u^2$. To streamline notation and for clarity of presentation, we can rewrite the adaptive law (4.39) as

$$\dot{\theta} = \gamma \epsilon u, \ \theta(0) = \theta_0 \tag{4.40}$$

where

$$\epsilon := \frac{\epsilon_1}{m^2} \tag{4.41}$$

is called the *normalized* estimation error. The adaptive law (4.40) which uses the normalized estimation error is called the *normalized gradient* adaptive law as opposed to the adaptive law (4.6) which is called the *unnormalized gradient* adaptive law since it uses the unnormalized estimation error ϵ_1. Next we analyze the properties of the normalized adaptive law (4.40).

As in the last section, define $\tilde{\theta} = \theta - \theta^*$ and consider the Lyapunov-like function $V(\tilde{\theta}) = \frac{\tilde{\theta}^2}{2\gamma}$. Then along the solution of (4.40) we have

$$\begin{aligned} \dot{V} &= \epsilon \tilde{\theta} u \\ &= \epsilon(-\epsilon m^2) \text{ (using (4.41) and the fact that } \epsilon_1 = y - \theta u = -\tilde{\theta} u) \\ &= -\epsilon^2 m^2 \\ &\leq 0 \end{aligned}$$

Hence $\tilde{\theta}, \theta$ are bounded and $\epsilon m \in L_2$. Since $\tilde{\theta}, \frac{u}{m} \in L_\infty$, it follows from $\epsilon = -\frac{\tilde{\theta} u}{m^2}$ that $\epsilon, \epsilon m \in L_\infty$. Now $\dot{\theta} = \gamma \epsilon m \frac{u}{m}$. Since $\frac{u}{m} \in L_\infty$ and $\epsilon m \in L_2 \cap L_\infty$, it follows that $\dot{\theta} \in L_2 \cap L_\infty$. All of the above properties are established without imposing any restrictions on u.

Now $\frac{d}{dt}(\epsilon m) = \frac{d}{dt}(-\tilde{\theta}\bar{u}) = -\dot{\tilde{\theta}}\bar{u} - \tilde{\theta}\dot{\bar{u}}$. Since $\dot{\tilde{\theta}}, \bar{u}, \tilde{\theta}$ are all bounded, it follows that if in addition $\dot{\bar{u}} \in L_\infty$, then we can use Corollary 2.2.1 to conclude that $\epsilon m(t) \to 0$ as $t \to \infty$, and this completes the analysis of the properties of (4.40).

Remark 4.4.1. When $n_s = 0$, i.e. $m = 1$, the adaptive law (4.40) becomes the unnormalized adaptive law (4.6) considered earlier. An important difference between the two adaptive laws is that while (4.6) does not guarantee $\dot{\theta} \in L_2$ unless u is assumed to be apriori bounded, the normalized adaptive law (4.40) does so without any additional assumptions on u. Such a property is not very important for parameter estimation purposes but plays a crucial role if the parameter estimates are used to design adaptive controllers. This will be clearly seen in the next chapter.

4.4.2 First Order Example

Consider the first order plant

$$\dot{y} = -ay + bu, \; y(0) = y_0 \tag{4.42}$$

where the parameters a and b are unknown and the signals $u(t)$ and $y(t)$ are not necessarily bounded. Our objective here is to design a parameter estimator for estimating the unknown parameters a and b. Although the input and output signal boundedness assumptions of Section 4.2.2 are no longer valid here, nevertheless let us examine how far we can proceed as in Section 4.2.2, before running into difficulties.

Filtering each side of (4.42) using the stable filter $\Lambda(s) = \frac{1}{s+1}$, we come up with the linear parametric models

$$z = \theta^{*T}\phi \tag{4.43}$$

where $\phi = \left[\frac{1}{s+1}[y], \frac{1}{s+1}[u]\right]^T$ and either $z = \frac{s}{s+1}[y], \theta^* = [-a, b]$ or $z = y$ and $\theta^* = [-a+1, b]^T$. In either case, let $\theta(t)$ be the estimate of θ^*. Then we can define the estimate $\hat{z}$ of z as

$$\hat{z} = \theta^T\phi \tag{4.44}$$

and use it to construct the estimation error

$$\epsilon_1 = z - \hat{z} = z - \theta^T\phi \tag{4.45}$$

If we now consider the instantaneous cost function $J = \frac{\epsilon_1^2}{2} = \frac{(z-\theta^T\phi)^2}{2}$ and try to minimize it with respect to θ, the result is an ill-posed minimization problem since ϕ and z are not guaranteed to be bounded. So, as in the last subsection, we divide each side of (4.43) by some function $m > 0$, called a normalizing signal, to obtain

$$\bar{z} = \theta^{*T}\bar{\phi} \tag{4.46}$$

where $\bar{z} = \frac{z}{m}$ and $\bar{\phi} = \frac{\phi}{m}$ and $m^2 = 1 + n_s^2$. The signal n_s is chosen so that $\bar{\phi} \in L_\infty$ i.e. $\frac{\phi}{m} \in L_\infty$, e.g. $n_s^2 = \phi^T\phi$. Since $\bar{\phi}$, and hence $\bar{z} \in L_\infty$, we can start with (4.46) and proceed as in Section 4.2.2 to obtain the normalized adaptive law:

Let $\theta(t)$ be the estimate of θ^* at time t. Then, in view of (4.46), we obtain the estimate $\hat{\bar{z}}$ of $\bar{z}$ as

$$\hat{\bar{z}} = \theta^T\bar{\phi} \tag{4.47}$$

Define the estimation error $\bar{\epsilon}_1$ by

$$\bar{\epsilon}_1 = \bar{z} - \hat{\bar{z}} = \bar{z} - \theta^T\bar{\phi} \tag{4.48}$$

and consider the instantaneous cost function $J = \frac{\bar{\epsilon}_1^2}{2} = \frac{(\bar{z}-\theta^T\bar{\phi})^2}{2} = \frac{(z-\theta^T\phi)^2}{2m^2}$. Since $\frac{\phi}{m}, \frac{z}{m} \in L_\infty$, it follows that J cannot be driven to ∞ by ϕ or z. Hence, minimizing J with respect to θ is a meaningful optimization problem, and using the gradient method, we obtain the normalized adaptive law

$$\dot{\theta} = \gamma \bar{\epsilon}_1 \bar{\phi}, \ \gamma > 0, \ \theta(0) = \theta_0 \tag{4.49}$$

Substituting $\bar{\epsilon}_1 = \frac{\epsilon_1}{m}$, $\bar{\phi} = \frac{\phi}{m}$, the above adaptive law becomes

$$\dot{\theta} = \frac{\gamma \epsilon_1 \phi}{m^2}, \ \theta(0) = \theta_0 \tag{4.50}$$

As in the last subsection, we can define the normalized estimation error ϵ as

$$\epsilon := \frac{\epsilon_1}{m^2} \tag{4.51}$$

so that the normalized adaptive law (4.50) can be rewritten as

$$\dot{\theta} = \gamma \epsilon \phi, \ \theta(0) = \theta_0 \tag{4.52}$$

Next we analyze the properties of the normalized adaptive law (4.52). As many times before, we define $\tilde{\theta} = \theta - \theta^*$ and consider the Lyapunov-like function $V(\tilde{\theta}) = \frac{\tilde{\theta}^T \tilde{\theta}}{2\gamma}$. Then, along the solution of (4.52), we have

$$\begin{aligned} \dot{V} &= \epsilon \tilde{\theta}^T \phi \\ &= \epsilon(-\epsilon m^2) \text{ (using (4.51) and the fact that } \epsilon_1 = z - \theta^T \phi = -\tilde{\theta}^T \phi) \\ &= -\epsilon^2 m^2 \\ &\leq 0 \end{aligned}$$

Hence, $\tilde{\theta}, \theta$ are bounded and $\epsilon m \in L_2$. Since $\tilde{\theta}, \frac{\phi}{m} \in L_\infty$, it follows from $\epsilon = -\frac{\tilde{\theta}^T \phi}{m^2}$ that $\epsilon, \epsilon m \in L_\infty$. Now $\dot{\theta} = \gamma \epsilon m \frac{\phi}{m}$. Since $\frac{\phi}{m} \in L_\infty$ and $\epsilon m \in L_2 \cap L_\infty$, it follows that $\dot{\theta} \in L_2 \cap L_\infty$. If in addition to $\frac{\phi}{m} \in L_\infty$, we also have $\frac{d}{dt}\left(\frac{\phi}{m}\right) \in L_\infty$, then $\frac{d}{dt}(\epsilon m) = \frac{d}{dt}(-\tilde{\theta}^T \bar{\phi}) = -\dot{\tilde{\theta}}^T \bar{\phi} - \tilde{\theta}^T \dot{\bar{\phi}}$ belongs to L_∞ and hence, we can use Corollary 2.2.1 to conclude that $\epsilon m \to 0$ as $t \to \infty$.

Remark 4.4.2. The normalized adaptive law (4.52) reduces to the unnormalized adaptive law (4.17) when n_s is chosen to be equal to zero i.e. $m = 1$. Also, as in the case of the scalar example, the normalized adaptive law (4.52) guarantees that $\dot{\theta} \in L_2$ independent of the boundedness of ϕ, while the corresponding unnormalized adaptive law (4.17) does not guarantee such a property unless ϕ is assumed to be apriori bounded.

4.4.3 General Plant

In this subsection, we consider the general linear time-invariant finite dimensional plant described by the input-output differential equation

$$y^{(n)} + a_1 y^{(n-1)} + a_2 y^{(n-2)} + \cdots + a_n y = b_0 u^{(l)} + b_1 u^{(l-1)} + \cdots + b_l u \quad (4.53)$$

where $n > l$ and $u(t), y(t)$ represent the input and output signals of the plant respectively. The parameter estimation problem here is similar to the one considered in Section 4.3 except that the signals u and y are no longer necessarily bounded. Let us now use the approach developed in Section 4.3 and see how far we can proceed before encountering difficulties. As in Section 4.3, and for reasons outlined therein, we filter both sides of (4.53) with $\frac{1}{\Lambda(s)}$, where $\Lambda(s) = s^n + \lambda_1 s^{n-1} + \cdots + \lambda_n$ is an arbitrary monic Hurwitz polynomial of degree n. This leads to the plant representation

$$z = \theta^{*T} \phi \quad (4.54)$$

$$\begin{aligned} \text{where } \phi &= \left[\frac{s^{n-1}}{\Lambda(s)}[y], \frac{s^{n-2}}{\Lambda(s)}[y], \cdots, \frac{1}{\Lambda(s)}[y], \frac{s^l}{\Lambda(s)}[u], \frac{s^{l-1}}{\Lambda(s)}[u],\right. \\ &\quad \left. \cdots, \frac{1}{\Lambda(s)}[u]\right]^T && (4.55) \\ \text{and either } z &= \frac{s^n}{\Lambda(s)}[y], && (4.56) \\ \theta^* &= [-a_1, -a_2, \cdots, -a_n, b_0, b_1, \cdots, b_l]^T && (4.57) \\ \text{or } z &= y, && (4.58) \\ \theta^* &= [\lambda_1 - a_1, \lambda_2 - a_2, \cdots, \lambda_n - a_n, b_0, b_1, \cdots, b_l]^T && (4.59) \end{aligned}$$

Now, let $\theta(t)$ be the estimate of θ^* at time t. Then, in view of (4.54), the estimate $\hat{z}$ of z can be constructed as

$$\hat{z} = \theta^T \phi \quad (4.60)$$

based on which we can define the estimation error

$$\epsilon_1 = z - \hat{z} = z - \theta^T \phi \quad (4.61)$$

If we now consider the instantaneous cost function $J = \frac{\epsilon_1^2}{2} = \frac{(z - \theta^T \phi)^2}{2}$ and try to minimize it with respect to θ, then we have an ill-posed minimization problem since ϕ and z are not necessarily bounded and could possibly drive J to infinity. As in the case of the scalar and first order examples, we remedy the situation by dividing each side of (4.54) by some function $m > 0$, called a normalizing signal, to obtain

$$\bar{z} = \theta^{*T} \bar{\phi} \quad (4.62)$$

where $\bar{z} = \frac{z}{m}$ and $\bar{\phi} = \frac{\phi}{m}$ and $m^2 = 1 + n_s^2$. The signal n_s is chosen so that $\bar{\phi} \in L_\infty$, i.e. $\frac{\phi}{m} \in L_\infty$. A possible choice for n_s could be $n_s^2 = \phi^T\phi$. Since $\bar{\phi}$, and hence $\bar{z} \in L_\infty$, we can start with (4.62) and proceed as in Section 4.3 to obtain the normalized adaptive law:

Let $\theta(t)$ be the estimate of θ^* at time t. Then, in view of (4.62), we obtain the estimate $\hat{\bar{z}}$ of $\bar{z}$ as

$$\hat{\bar{z}} = \theta^T \bar{\phi} \tag{4.63}$$

Define the estimation error $\bar{\epsilon}_1$ by

$$\bar{\epsilon}_1 = \bar{z} - \hat{\bar{z}} = \bar{z} - \theta^T \bar{\phi} \tag{4.64}$$

and consider the instantaneous cost function $J = \frac{\bar{\epsilon}_1^2}{2} = \frac{(\bar{z}-\theta^T\bar{\phi})^2}{2} = \frac{(z-\theta^T\phi)^2}{2m^2}$. Since $\frac{\phi}{m}, \frac{z}{m} \in L_\infty$, it follows that J cannot be driven to infinity by ϕ or z. Hence, minimizing J with respect to θ is a meaningful optimization problem, and using the gradient method, we obtain the normalized adaptive law

$$\dot{\theta} = \gamma \bar{\epsilon}_1 \bar{\phi}, \ \gamma > 0, \ \theta(0) = \theta_0 \tag{4.65}$$

Substituting $\bar{\epsilon}_1 = \frac{\epsilon_1}{m}, \bar{\phi} = \frac{\phi}{m}$, the above adaptive law becomes

$$\dot{\theta} = \frac{\gamma \epsilon_1 \phi}{m^2}, \ \theta(0) = \theta_0 \tag{4.66}$$

Defining the normalized estimation error ϵ as

$$\epsilon = \frac{\epsilon_1}{m^2} = \frac{z - \theta^T\phi}{m^2} \tag{4.67}$$

the normalized adaptive law becomes

$$\dot{\theta} = \gamma \epsilon \phi, \ \theta(0) = \theta_0 \tag{4.68}$$

The following Theorem summarizes the properties of the normalized adaptive law (4.68).

Theorem 4.4.1. *Consider the normalized gradient update law (4.68) based on the linear parametric model (4.54) where the signal vector ϕ is not necessarily bounded i.e. ϕ may or may not be bounded. Then the following properties are guaranteed: (i) $\theta \in L_\infty$ and (ii) $\epsilon, \epsilon n_s, \dot{\theta} \in L_2 \cap L_\infty$. If in addition $\frac{d}{dt}\left(\frac{\phi}{m}\right) \in L_\infty$, then $\epsilon m, \epsilon \to 0$ as $t \to \infty$.*

Proof. Define the parameter error $\tilde{\theta} \triangleq \theta - \theta^*$ and consider the Lyapunov-like function $V(\tilde{\theta}) = \frac{\tilde{\theta}^T\tilde{\theta}}{2\gamma}$. Then, along the solution of (4.68), we have

$$\begin{aligned} \dot{V} &= \tilde{\theta}^T \epsilon \phi \\ &= -\epsilon^2 m^2 \text{ (using (4.54) and (4.67))} \\ &\leq 0 \end{aligned} \tag{4.69}$$

Hence $\tilde{\theta}, \theta$ are bounded and $\epsilon m \in L_2$. Since $\epsilon = -\frac{\tilde{\theta}^T \phi}{m^2}$ and $\frac{\phi}{m} \in L_\infty$, it follows that $\epsilon m \in L_\infty$ which in turn implies that $\epsilon, \epsilon n_s \in L_\infty$. Also $\epsilon^2 m^2 = \epsilon^2 + \epsilon^2 n_s^2$. Hence $\epsilon m \in L_2$ implies that $\epsilon, \epsilon n_s \in L_2$. Now $\dot{\theta} = \gamma \epsilon \phi = \gamma \epsilon m \frac{\phi}{m}$. Since $\frac{\phi}{m} \in L_\infty$ and $\epsilon m \in L_2 \cap L_\infty$, it follows that $\dot{\theta} \in L_2 \cap L_\infty$. The above properties are established without making any assumptions on ϕ. Let us now assume that $\frac{d}{dt}\left(\frac{\phi}{m}\right) \in L_\infty$. Now from (4.54), (4.67), we have $\epsilon = -\frac{\tilde{\theta}^T \phi}{m^2}$ so that $\epsilon m = -\tilde{\theta}^T \bar{\phi}$. Hence $\frac{d}{dt}(\epsilon m) = -\dot{\tilde{\theta}}^T \bar{\phi} - \tilde{\theta}^T \dot{\bar{\phi}}$. Since $\dot{\tilde{\theta}}, \bar{\phi}, \tilde{\theta} \in L_\infty$, it follows that if $\dot{\bar{\phi}} \in L_\infty$ then $\frac{d}{dt}(\epsilon m) \in L_\infty$ and by using Corollary 2.2.1, we conclude that $\lim_{t \to \infty} \epsilon m(t) = 0$. Furthermore, since $m \geq 1$, it follows that $\lim_{t \to \infty} \epsilon(t) = 0$, and the proof is complete.

Remark 4.4.3. As in the case of the simple examples, the normalized adaptive law (4.68) reduces to the corresponding unnormalized adaptive law (4.32) when n_s is chosen to be equal to zero i.e. $m = 1$. Also, the normalized adaptive law (4.68) guarantees that $\dot{\theta} \in L_2$ independent of the boundedness of ϕ, while the unnormalized adaptive law (4.32) does not guarantee such a property unless ϕ is assumed to be apriori bounded.

4.5 Adaptive Laws with Projection

In many parameter estimation problems, one may have some apriori knowledge about the unknown parameters being estimated. This knowledge may come from known physical constraints on the values that the unknown parameters can take. For instance, if the unknown parameter being estimated is the mass of a physical system, it immediately follows that such a parameter cannot assume negative values and so it would make sense to constrain the parameter estimates to the positive real axis. The gradient adaptive laws designed so far are by themselves not capable of guaranteeing such a property. Such a property can, however, be guaranteed by modifying the original gradient descent algorithms by using the so called *Gradient Projection Method.* As we will see in this section, a remarkable feature of these modified algorithms is that they are able to retain the properties established earlier for the original adaptive laws.

Suppose that the unknown parameter vector θ^* in the linear parametric model

$$z = \theta^{*T} \phi \tag{4.70}$$

is apriori known to belong to $\mathcal{C}$, where $\mathcal{C}$ is a *convex set* in R^{n+l+1}. Furthermore, suppose that the set $\mathcal{C}$ is described as $\mathcal{C} = \left\{\theta \in R^{n+l+1} | g(\theta) \leq 0\right\}$ where $g : R^{n+l+1} \mapsto R$ is a smooth function. Then, one would start the parameter estimation process from inside the set $\mathcal{C}$ and make sure that the parameter estimate does not leave $\mathcal{C}$ as the estimation proceeds. In the case of the gradient method, one would now minimize J *subject to the constraint* $\theta \in \mathcal{C}$ leading to the following algorithm (see Appendix B, Equation (B.6)):

$$\dot{\theta} = \begin{cases} -\gamma\nabla J & \text{if } \theta \in \mathcal{C}^0 \text{ or if } \theta \in \delta\mathcal{C} \text{ and } \\ & -(\gamma\nabla J)^T\nabla g \leq 0 \\ -\gamma\nabla J + \frac{(\nabla g \nabla g^T)}{\nabla g^T \nabla g}\gamma\nabla J & \text{otherwise} \end{cases} \tag{4.71}$$

For instance, the algorithm (4.68) would be modified to

$$\dot{\theta} = \begin{cases} \gamma\epsilon\phi & \text{if } \theta \in \mathcal{C}^0 \text{ or if } \theta \in \delta\mathcal{C} \text{ and } \\ & (\gamma\epsilon\phi)^T\nabla g \leq 0 \\ \gamma\epsilon\phi - \frac{(\nabla g \nabla g^T)}{\nabla g^T \nabla g}\gamma\epsilon\phi & \text{otherwise} \end{cases} \tag{4.72}$$

The following Theorem describes the effect of the projection modification given by (4.71).

Theorem 4.5.1. *The gradient adaptive laws considered in this chapter with the projection modification given by (4.71) retain all their properties that were established in the absence of projection and, in addition, guarantee that* $\theta \in \mathcal{C}\ \forall\, t \geq 0$ *provided* $\theta(0) = \theta_0 \in \mathcal{C}$ *and* $\theta^* \in \mathcal{C}$.

Proof. Now from (4.71) we see that whenever $\theta \in \delta\mathcal{C}$, we have $\dot{\theta}^T\nabla g \leq 0$ which implies that the vector $\dot{\theta}$ points either inside $\mathcal{C}$ or along the tangent plane of $\delta\mathcal{C}$ at θ. Because $\theta(0) = \theta_0 \in \mathcal{C}$, it follows that $\theta(t)$ will never leave $\mathcal{C}$, i.e. $\theta(t) \in \mathcal{C}\ \forall\, t \geq 0$. The adaptive law (4.71) has the same form as the one without projection except for the additional term

$$Q = \begin{cases} \frac{\nabla g \nabla g^T}{\nabla g^T \nabla g}\gamma\nabla J & \text{if } \theta \in \delta\mathcal{C} \text{ and } -(\gamma\nabla J)^T\nabla g > 0 \\ 0 & \text{otherwise} \end{cases}$$

in the expression for $\dot{\theta}$. If we use the same function $V(\tilde{\theta}) = \frac{\tilde{\theta}^T\tilde{\theta}}{2\gamma}$ as in the unconstrained case to analyze the adaptive law with projection, the time derivative $\dot{V}$ of V will have the additional term

$$\frac{\tilde{\theta}^T Q}{\gamma} = \begin{cases} \tilde{\theta}^T \frac{\nabla g \nabla g^T}{\nabla g^T \nabla g}\nabla J & \text{if } \theta \in \delta\mathcal{C} \text{ and } -(\gamma\nabla J)^T\nabla g > 0 \\ 0 & \text{otherwise} \end{cases}$$

Because of the convexity of $\mathcal{C}$ and the assumption that $\theta^* \in \mathcal{C}$, we have $\tilde{\theta}^T\nabla g = (\theta - \theta^*)^T\nabla g \geq 0$ when $\theta \in \delta\mathcal{C}$. Since $\nabla g^T\gamma\nabla J = (\gamma\nabla J)^T\nabla g < 0$ for $\theta \in \delta\mathcal{C}$ and $-(\gamma\nabla J)^T\nabla g > 0$, it follows that $\frac{\tilde{\theta}^T Q}{\gamma} \leq 0$. Therefore, the term $\frac{\tilde{\theta}^T Q}{\gamma}$ introduced by the projection can only make $\dot{V}$ more negative and does not affect the results developed from the properties of V and $\dot{V}$. Furthermore, the L_2 properties of $\dot{\theta}$ will not be affected by the projection because, with or without projection, $\dot{\theta}$ can be shown to satisfy $|\dot{\theta}|^2 \leq c|\gamma\nabla J|^2$ for some constant $c \in R^+$.

We now present two simple examples to illustrate parameter projection using the gradient projection method.

Example 4.5.1. Consider the scalar parameter estimation problem of Section 4.2.1, i.e. we are interested in estimating the scalar parameter θ^* from measurements of $y(t)$ and $u(t)$ where $y(t)$, $u(t)$ and θ^* are related by

$$y(t) = \theta^* u(t) \tag{4.73}$$

Furthermore, suppose that we know apriori that $|\theta^*| \leq c$ where $c > 0$ is a given constant. Thus, it would make sense to start with an initial parameter estimate satisfying this constraint and then make sure that the parameter estimate continues to satisfy this constraint as the parameter estimation procedure evolves.

In order to use the gradient projection method, we first observe that, in this case, the set $\mathcal{C} = \{\theta \in R| \ |\theta| \leq c\}$ which can also be described as $\mathcal{C} = \{\theta \in R | g(\theta) \leq 0\}$ where $g(\theta) = \theta^2 - c^2$. Then $\nabla g(\theta) = 2\theta$ and by using (4.71), the algorithm (4.40) becomes

$$\dot{\theta} = \begin{cases} \gamma\epsilon u & \text{if } |\theta| < c \text{ or if } |\theta| = c \text{ and } (\gamma\epsilon u)2\theta \leq 0 \\ \gamma\epsilon u - \frac{(2\theta)(2\theta)^T}{(2\theta)^2}\gamma\epsilon u & \text{otherwise} \end{cases}$$

$$\text{or } \dot{\theta} = \begin{cases} \gamma\epsilon u & \text{if } |\theta| < c \text{ or if } |\theta| = c \text{ and } (\gamma\epsilon u)2\theta \leq 0 \\ 0 & \text{otherwise} \end{cases} \tag{4.74}$$

In view of Theorem 4.5.1, the adaptive law (4.74) ensures that $|\theta(t)| \leq c \, \forall \, t \geq 0$ provided $|\theta(0)| \leq c$. To do so, it does something which is intuitively obvious — if $\theta(t)$ reaches the boundary of the set $\mathcal{C}$, it continues with the unconstrained adaptation algorithm (4.40) as long as $\theta(t)$ is not trying to exit $\mathcal{C}$ i.e. $\theta\dot{\theta} \leq 0$; otherwise, it freezes the adaptation to keep θ inside the set $\mathcal{C}$.

Example 4.5.2. Consider the parameter estimation problem of Section 4.2.2, i.e. we are interested in estimating the parameters a and b of the first order stable plant

$$\dot{y} = -ay + bu, \; y(0) = y_0 \tag{4.75}$$

where $a > 0$ and the signals $u(t)$ and $y(t)$ are available for measurement. As shown in Section 4.2.2, with $\theta^* \triangleq [-a, b]^T$, $\phi \triangleq \left[\frac{1}{s+1}[y], \frac{1}{s+1}[u]\right]^T$, $z \triangleq \frac{s}{s+1}[y]$, we can design the unconstrained gradient algorithm

$$\dot{\theta} = \gamma\epsilon_1\phi, \; \gamma > 0, \; \theta(0) = \theta_0 \tag{4.76}$$

$$\epsilon_1 = z - \theta^T\phi \tag{4.77}$$

We would now like to modify the adaptive law (4.76) so that the parameter estimate θ belongs to a convex set $\mathcal{C}$ in R^2. Furthermore, suppose that the convex set $\mathcal{C}$ that we are interested in is given by $\mathcal{C} = \{\theta \in R^2 | \theta_i \in [\theta_{il}, \theta_{iu}],\ i = 1, 2\}$, i.e. $\mathcal{C}$ is the Cartesian product of intervals $[\theta_{1l}, \theta_{1u}] \times [\theta_{2l}, \theta_{2u}]$. This set $\mathcal{C}$ can be descibed as the intersection of four convex sets $\mathcal{C}_1$, $\mathcal{C}_2$, $\mathcal{C}_3$, $\mathcal{C}_4$ where

$$\begin{aligned} \mathcal{C}_1 &= \{\theta \in R^2 | \theta_1 \geq \theta_{1l}\} \\ \mathcal{C}_2 &= \{\theta \in R^2 | \theta_1 \leq \theta_{1u}\} \\ \mathcal{C}_3 &= \{\theta \in R^2 | \theta_2 \geq \theta_{2l}\} \\ \mathcal{C}_4 &= \{\theta \in R^2 | \theta_2 \leq \theta_{2u}\} \end{aligned}$$

The functions g_i, $i = 1, 2, 3, 4$ describing each of these sets is given by $g_1(\theta) = \theta_{1l} - \theta_1$, $g_2(\theta) = \theta_1 - \theta_{1u}$, $g_3(\theta) = \theta_{2l} - \theta_2$, $g_4(\theta) = \theta_2 - \theta_{2u}$ so that $\nabla g_1(\theta) = [-1, 0]^T$, $\nabla g_2(\theta) = [1, 0]^T$, $\nabla g_3(\theta) = [0, -1]^T$, $\nabla g_4(\theta) = [0, 1]^T$.

We now proceed to apply the gradient projection technique to derive an adaptive law which guarantess that $\theta(t) \in \mathcal{C}\ \forall t \geq 0$. Since we have four convex sets here, we will have to treat them in succession. In order to be able to do so, we first make the following simple observation based on (4.5) and (4.71).

Suppose that prior to implementing a parameter projection, $\dot{\theta}$ is given by the adaptive law

$$\dot{\theta} = f$$

where f is a function of time and other signals. Then, when parameter projection, as in the gradient projection method, is implemented, the new expression for $\dot{\theta}$ becomes

$$\dot{\theta} = \begin{cases} f & \text{if } \theta \in \mathcal{C}^0 \text{ or if } \theta \in \delta\mathcal{C} \text{ and} \\ & f^T \nabla g \leq 0 \\ f - \frac{\nabla g \nabla g^T}{\nabla g^T \nabla g} f & \text{otherwise} \end{cases} \tag{4.78}$$

It can be easily verified as in the proof of Theorem 4.5.1 that such a projection will not affect the results developed from the properties of $V = \frac{(\theta - \theta^*)^T (\theta - \theta^*)}{2\gamma}$ and its derivative, and will also ensure that $\theta(t) \in \mathcal{C}$ provided $\theta(0) \in \mathcal{C}$. For clarity of the presentation to follow, let us rewrite (4.78) as

$$\dot{\theta} = \begin{cases} f - \frac{\nabla g \nabla g^T}{\nabla g^T \nabla g} f & \text{if } \theta \in \delta\mathcal{C} \text{ and } f^T \nabla g > 0 \\ f & \text{otherwise} \end{cases} \tag{4.79}$$

We now start from the unconstrained gradient algorithm (4.76) and successively use (4.79) with $\mathcal{C} = \mathcal{C}_1, \mathcal{C}_2, \mathcal{C}_3$ and $\mathcal{C}_4$. First, with $\mathcal{C} = \mathcal{C}_1$, we obtain

$$\dot{\theta} = Pr_1[\gamma\epsilon_1\phi] = \begin{cases} \gamma\epsilon_1\phi - \begin{bmatrix} 1 & 0 \\ 0 & 0 \end{bmatrix} \gamma\epsilon_1\phi & \text{if } \theta \in \delta\mathcal{C}_1 \text{ and} \\ & \gamma\epsilon_1\phi^T \begin{bmatrix} -1 \\ 0 \end{bmatrix} > 0 \\ \gamma\epsilon_1\phi & \text{otherwise} \end{cases}$$

$$= \begin{cases} \gamma\epsilon_1 \begin{bmatrix} 0 \\ \phi_2 \end{bmatrix} & \text{if } \theta_1 = \theta_{1l} \text{ and} \\ & \gamma\epsilon_1\phi_1 < 0 \\ \gamma\epsilon_1\phi & \text{otherwise} \end{cases} \tag{4.80}$$

The algorithm (4.80) agrees with our intuition: if θ_1 reaches the lower limit and is decreasing, freeze adaptation in that component; otherwise, continue with the unconstrained adaptation.

Next with $\mathcal{C} = \mathcal{C}_2$, we obtain

$$\begin{aligned} \dot{\theta} &= Pr_2[Pr_1[\gamma\epsilon_1\phi]] \\ &= \begin{cases} Pr_1[\gamma\epsilon_1\phi] - \begin{bmatrix} 1 & 0 \\ 0 & 0 \end{bmatrix} Pr_1[\gamma\epsilon_1\phi] & \text{if } \theta \in \delta\mathcal{C}_2 \text{ and} \\ & (Pr_1[\gamma\epsilon_1\phi])^T \begin{bmatrix} 1 \\ 0 \end{bmatrix} > 0 \\ Pr_1[\gamma\epsilon_1\phi] & \text{otherwise} \end{cases} \end{aligned}$$

which again says that we should continue with $\dot{\theta} = Pr_1[\gamma\epsilon_1\phi]$ unless $\theta_1 = \theta_{1u}$ and $\gamma\epsilon_1\phi_1 > 0$, i.e. θ_1 reaches the upper limit and is trying to increase, in which case the adaptation in that component should be shut off. Thus the net result is that after projecting on $\mathcal{C}_1$ and $\mathcal{C}_2$, we have

$$\dot{\theta} = Pr_2[Pr_1[\gamma\epsilon_1\phi]] = \begin{cases} \gamma\epsilon_1 \begin{bmatrix} 0 \\ \phi_2 \end{bmatrix} & \text{if } \theta_1 = \theta_{1l} \text{ and } \gamma\epsilon_1\phi_1 < 0 \\ & \text{or if } \theta_1 = \theta_{1u} \text{ and } \gamma\epsilon_1\phi_1 > 0 \\ \gamma\epsilon_1\phi & \text{otherwise} \end{cases}$$

We note that the projection onto the convex sets $\mathcal{C}_1$ and $\mathcal{C}_2$ affects only the first component of $\dot{\theta}$. Similarly, we can show that by now projecting onto $\mathcal{C}_3$ and then $\mathcal{C}_4$, only the second component of $\dot{\theta}$ will be affected. The detailed steps are straightforward but notationally cumbersome and are left to the reader. The final result is that we come up with the adaptive law

$$\dot{\theta}_1 = \begin{cases} 0 & \text{if } \theta_1 = \theta_{1l} \text{ and } \gamma\epsilon_1\phi_1 < 0 \\ & \text{or if } \theta_1 = \theta_{1u} \text{ and } \gamma\epsilon_1\phi_1 > 0 \\ \gamma\epsilon_1\phi_1 & \text{otherwise} \end{cases}$$

$$\dot{\theta}_2 = \begin{cases} 0 & \text{if } \theta_2 = \theta_{2l} \text{ and } \gamma\epsilon_1\phi_2 < 0 \\ & \text{or if } \theta_2 = \theta_{2u} \text{ and } \gamma\epsilon_1\phi_2 > 0 \\ \gamma\epsilon_1\phi_2 & \text{otherwise} \end{cases}$$

Of course, the initial parameter estimates must be chosen to belong to $\mathcal{C}$,i.e. $\theta_1(0) \in [\theta_{1l}, \theta_{1u}]$ and $\theta_2(0) \in [\theta_{2l}, \theta_{2u}]$. We further observe that if $\theta_{2l}\theta_{2u} > 0$, then the above algorithm will guarantee that the estimate of b does not pass through zero, a property which is important for some of the adaptive IMC schemes to be discussed in this monograph.

Remark 4.5.1. In the adaptive laws considered in this chapter, the adaptive gain was always chosen to be γ for each component of the parameter vector. In other words, the adaptive gain matrix is γI where I is the identity matrix. This is only for the sake of simplicity. Indeed, similar results can be obtained by choosing the adaptive gain matrix Γ to be any positive definite, symmetric matrix. The interested reader is referred to [19] for the details.

CHAPTER 5
ADAPTIVE INTERNAL MODEL CONTROL SCHEMES

5.1 Introduction

In the last two chapters, we have studied internal model control schemes and parameter estimators as two separate entities. Chapter 3 dealt with the internal model control of stable plants with ***known parameters*** while in Chapter 4 we developed on-line parameter estimation techniques for estimating the unknown parameters of a given plant. In this chapter, our main objective is to design internal model controllers for stable plants with ***unknown parameters***. The intuitively obvious way of achieving such an objective is to design an on-line parameter estimator for estimating the unknown plant parameters as in Chapter 4, and then to use the techniques of Chapter 3 to design an internal model controller based on these parameter estimates. This approach of treating the estimated parameters as the true ones, and basing the control design on them is referred to in the adaptive literature as *Certainty Equivalence.* Although the estimated parameters in a certainty equivalence scheme rarely converge to the true values, nevertheless research in adaptive control theory over the last two decades has shown that many designs based on the certainty equivalence approach can be ***proven to be stable*** [19, 31, 32]. Unfortunately, adaptive internal model control (AIMC) schemes were not included in this category, presumably because they arose in the context of industrial applications and consequently did not attract much attention from the theoreticians. Indeed, the literature on Adaptive Internal Model Control is replete with simulations and empirical studies showing the efficacy of certainty equivalence based adaptive internal model control schemes but hardly any instance exists where theoretical guarantees of stability and/or performance were obtained [40, 39]. Thus an important aspect of our treatment in this chapter will be the ***provable*** guarantees of stability, performance, etc. provided by the AIMC schemes to be designed.

For clarity of presentation, our treatment in this chapter will focus on the ideal case, i.e. we will assume for the time being that the plant to be controlled is a perfectly modelled one. Furthermore, let the plant be described by the input-output relationship

$$y = P(s)[u] = \frac{Z_0(s)}{R_0(s)}[u] \tag{5.1}$$

where u, y are the input and output signals of the plant; $P(s) = \frac{Z_0(s)}{R_0(s)}$ is the plant transfer function; $R_0(s)$ is a monic Hurwitz polynomial of degree n; $Z_0(s)$ is a polynomial of degree l with $l < n$; and the coefficients of $Z_0(s)$ and $R_0(s)$ are unknown. In order to implement a certainty equivalence based AIMC scheme, we need to first design a parameter estimator to provide on-line estimates of the unknown parameters. This can be done using the results in Chapter 4.

5.2 Design of the Parameter Estimator

Let $\Lambda(s)$ be an arbitrary monic Hurwitz polynomial of degree n. Then, as in Chapter 4, the plant equation (5.1) can be rewritten as

$$y = \theta^{*^T} \phi \tag{5.2}$$

where $\theta^* = [\theta_1^{*^T}, \theta_2^{*^T}]^T$; θ_1^*, θ_2^* are vectors containing the coefficients of $[\Lambda(s) - R_0(s)]$ and $Z_0(s)$ respectively;

$$\begin{aligned} \phi &= [\phi_1^T, \phi_2^T]^T; \\ \phi_1 &= \frac{a_{n-1}(s)}{\Lambda(s)}[y], \phi_2 = \frac{a_l(s)}{\Lambda(s)}[u]; \\ a_{n-1}(s) &= \left[s^{n-1}, s^{n-2}, \ldots, 1\right]^T \\ \text{and } a_l(s) &= \left[s^l, s^{l-1}, \ldots, 1\right]^T. \end{aligned} \tag{5.3}$$

Since the parameter estimates are to be used for *control design* purposes, we cannot apriori assume the boundedness of the signal vector ϕ. Furthermore, depending on the particular IMC scheme being considered, the "estimated plant" based on the parameter estimates will have to satisfy certain properties: for instance, for an adaptive model reference control scheme, the estimated plant will have to be pointwise minimum-phase. These specific requirements will vary from one controller to the other and will, therefore, be discussed in detail in the next section when we consider the actual control designs. Nevertheless, since the estimated plant will have to satisfy these properties, it is imperative that such a feature be built into the parameter estimator. To do so, we assume that convex sets C_θ, which will differ from one IMC scheme to the other, are known in the parameter space such that for every θ in C_θ, the corresponding plant satisfies the desired properties. Then by projecting θ onto C_θ, we can ensure that the parameter estimates do have the desired properties required for a particular control design. Using the results of Sections 4.4 and 4.5 of the last chapter, we obtain the following *normalized gradient* adaptive law with *parameter projection*:

$$\dot{\theta} = Pr[\gamma\varepsilon\phi], \ \theta(0) = \theta_0 \in C_\theta \tag{5.4}$$

$$\varepsilon = \frac{y - \hat{y}}{m^2} \tag{5.5}$$

$$\hat{y} = \theta^T \phi \tag{5.6}$$

$$m^2 = 1 + n_s^2,\ n_s^2 = \phi^T \phi \tag{5.7}$$

where $\gamma > 0$ is an adaptive gain; $\mathcal{C}_\theta$ is a known compact[1] convex set containing θ^*; $Pr[\cdot]$ is the projection operator given by (4.72) which guarantees that the parameter estimate $\theta(t)$ does not exit the set $\mathcal{C}_\theta$.

Various adaptive IMC schemes can now be obtained by replacing the internal model in Fig. 3.5 by that obtained from Equation (5.6), and the IMC parameters $Q(s)$ by time-varying operators which implement the certainty equivalence versions of the controller structures considered in Chapter 3. The detailed design of these Certainty Equivalence controllers is discussed next.

5.3 Certainty Equivalence Control Laws

We first outline the steps involved in designing a general Certainty Equivalence Adaptive IMC scheme. Thereafter, additional simplifications or complexities that result from the use of a particular control law will be discussed.

- **Step 1:** First use the parameter estimate $\theta(t)$ obtained from the adaptive law (5.4)-(5.7) to generate estimates of the numerator and denominator polynomials of the plant[2]:

$$\begin{aligned} \hat{Z}_0(s,t) &= \theta_2^T(t) a_l(s) \\ \hat{R}_0(s,t) &= \Lambda(s) - \theta_1^T(t) a_{n-1}(s) \end{aligned}$$

- **Step 2:** Using the frozen time plant $\hat{P}(s,t) = \frac{\hat{Z}_0(s,t)}{\hat{R}_0(s,t)}$, calculate the appropriate $\hat{Q}(s,t)$ using the results developed in Chapter 3.
- **Step 3:** Express $\hat{Q}(s,t)$ as $\hat{Q}(s,t) = \frac{\hat{Q}_n(s,t)}{\hat{Q}_d(s,t)}$ where $\hat{Q}_n(s,t)$ and $\hat{Q}_d(s,t)$ are time-varying polynomials with $\hat{Q}_d(s,t)$ *being monic.*
- **Step 4:** Choose $\Lambda_1(s)$ to be an arbitrary monic Hurwitz polynomial of degree equal to that of $\hat{Q}_d(s,t)$, and let this degree be denoted by n_d.
- **Step 5:** In view of the equivalent IMC representation (3.5) introduced in Chapter 3, the certainty equivalence control law becomes

$$u = q_d^T(t) \frac{a_{n_d-1}(s)}{\Lambda_1(s)}[u] + q_n^T(t) \frac{a_{n_d}(s)}{\Lambda_1(s)}[r - \epsilon m^2] \tag{5.8}$$

---

[1] As will be seen later on in this chapter, the compactness assumption about $\mathcal{C}_\theta$ is inserted here only to aid in the stability analysis of the adaptive IMC schemes.

[2] In the rest of this chapter, the "hats" denote the time varying polynomials/frozen time "transfer functions" that result from replacing the time-invariant coefficients of a "hat-free" polynomial/transfer function by their corresponding time-varying values obtained from adaptation and/or certainty equivalence control.

where $q_d(t)$ is the vector of coefficients of $\Lambda_1(s)-\hat{Q}_d(s,t)$; $q_n(t)$ is the vector of coefficients of $\hat{Q}_n(s,t)$; $a_{n_d}(s) = [s^{n_d}, s^{n_d-1}, \cdots, 1]^T$ and $a_{n_d-1}(s) = [s^{n_d-1}, s^{n_d-2}, \cdots, 1]^T$.

The adaptive IMC scheme resulting from combining the control law (5.8) with the adaptive law (5.4)-(5.7) is schematically depicted in Fig. 5.1. We

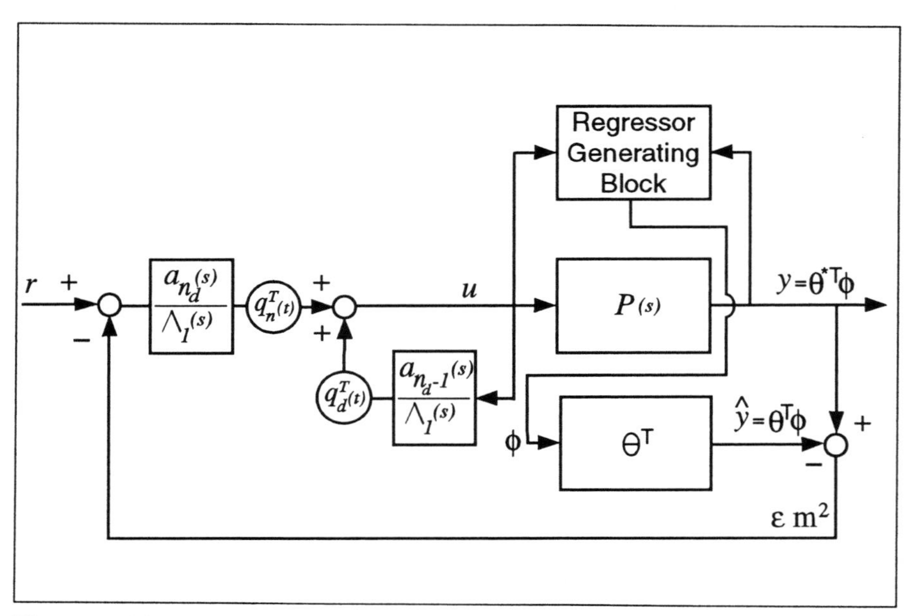

Fig. 5.1. Adaptive IMC Scheme

now proceed to discuss the simplifications or additional complexities that result from the use of each of the controller structures presented in Chapter 3.

5.3.1 Adaptive Partial Pole Placement

In this case, the design of the IMC parameter does not depend on the estimated plant. Indeed, $Q(s)$ is a fixed stable transfer function and not a time varying operator. Since the estimated plant is not used for on-line control design purposes, the parameter estimates do not have to satisfy any special properties. As a result, in this case, no parameter projection is required in the adaptive law.

5.3.2 Adaptive Model Reference Control

In this case from (3.6), we see that the $\hat{Q}(s,t)$ in Step 2 of the Certainty Equivalence design becomes

$$\hat{Q}(s,t) = W_m(s)\left[\hat{P}(s,t)\right]^{-1}. \tag{5.9}$$

Our stability analysis to be presented in Section 5.5 is based on results in the area of slowly time varying systems, in particular Theorem 2.4.9. In order for these results to be applicable, it is required that the operator $\hat{Q}(s,t)$ be pointwise stable (for each fixed t) and also that the degree of $\hat{Q}_d(s,t)$ in Step 3 of the Certainty Equivalence Design not change with time. These two requirements can be satisfied as follows:

- The pointwise stability of $\hat{Q}(s,t)$ can be guaranteed by ensuring that the frozen time estimated plant is minimum phase, i.e. $\hat{Z}_o(s,t)$ is Hurwitz stable for every fixed t. To guarantee such a property for $\hat{Z}_o(s,t)$, the projection set $\mathcal{C}_\theta$ in (5.4)-(5.7) is chosen so that $\forall\, \theta \in \mathcal{C}_\theta$, the corresponding $Z_0(s,\theta) = \theta_2^T a_l(s)$ is Hurwitz stable. By restricting $\mathcal{C}_\theta$ to be a subset of a Cartesian product of closed intervals, results from Parametric Robust Control [23, 1] can be used to construct such a $\mathcal{C}_\theta$. Also, when the projection set $\mathcal{C}_\theta$ cannot be specified as a single convex set, results from *hysteresis switching* using a finite number of convex sets [29] can be used.
- The degree of $\hat{Q}_d(s,t)$ can be rendered time invariant by ensuring that the leading coefficient of $\hat{Z}_o(s,t)$ is not allowed to pass through zero. As shown in Chapter 4 (Example 4.5.2), this feature can be built into the adaptive law by assuming some knowledge about the sign and a lower bound on the absolute value of the leading coefficient of $Z_0(s,\theta)$ and then implementing a parameter projection onto a convex set $\mathcal{C}_\theta$.

We will, therefore, assume that for IMC based model reference adaptive control, the set $\mathcal{C}_\theta$ has been suitably chosen to guarantee that the estimate $\theta(t)$ obtained from (5.4)-(5.7) actually satisfies both of the properties mentioned above.

5.3.3 Adaptive H_2 Optimal Control

In this case, $\hat{Q}(s,t)$ is obtained by replacing $P_M^{-1}(s)$, $B_P^{-1}(s)$ on the right hand side of (3.10) with $\hat{P}_M^{-1}(s,t)$, $\hat{B}_P^{-1}(s,t)$ where $\hat{P}_M(s,t)$ is the minimum phase portion of $\hat{P}(s,t)$and $\hat{B}_P(s,t)$ is the Blaschke product containing the open right-half plane zeros of $\hat{Z}_0(s,t)$. Thus $\hat{Q}(s,t)$ is given by

$$\hat{Q}(s,t) = \hat{P}_M^{-1}(s,t)R_M^{-1}(s)[\hat{B}_P^{-1}(s,t)R_M(s)]_* F(s) \tag{5.10}$$

where $[.]_*$ denotes that after a partial fraction expansion, the terms corresponding to the poles of $\hat{B}_P^{-1}(s,t)$ are removed, and $F(s)$ is an IMC Filter used to force $\hat{Q}(s,t)$ to be proper. As will be seen in Section 5.5, specifically Lemma 5.5.1, the degree of $\hat{Q}_d(s,t)$ in Step 3 of the Certainty Equivalence Design can be kept constant *using a single fixed* $F(s)$ provided the leading coefficient of $\hat{Z}_o(s,t)$ is not allowed to pass through zero. Additionally $\hat{Z}_0(s,t)$ should not have any zeros on the imaginary axis. A parameter projection, as in the case of model reference adaptive control, can be incorporated into the adaptive law (5.4)-(5.7) to guarantee both of these properties.

5.3.4 Adaptive H_∞ Optimal Control (one interpolation constraint)

In this case, $\hat{Q}(s,t)$ is obtained by replacing $P(s)$, b_1 on the right hand side of (3.13) with $\hat{P}(s,t)$, $\hat{b}_1$, i.e.

$$\hat{Q}(s,t) = \left[1 - \frac{W(\hat{b}_1)}{W(s)}\right] \hat{P}^{-1}(s,t)F(s) \tag{5.11}$$

where $\hat{b}_1$ is the open right half plane zero of $\hat{Z}_0(s,t)$ and $F(s)$ is the IMC Filter. Since (5.11) assumes the presence of only one open right half plane zero, the estimated polynomial $\hat{Z}_o(s,t)$ must have only one open right half plane zero and none on the imaginary axis. Additionally the leading coefficient of $\hat{Z}_o(s,t)$ should not be allowed to pass through zero so that the degree of $\hat{Q}_d(s,t)$ in Step 3 of the Certainty Equivalence Design can be kept fixed using a single fixed $F(s)$. Once again, both of these properties can be guaranteed by the adaptive law by appropriately choosing the set $\mathcal{C}_\theta$.

Remark 5.3.1. The actual construction of the sets $\mathcal{C}_\theta$ for adaptive model reference, adaptive H_2 and adaptive H_∞ optimal control may not be straight forward especially for higher order plants. However, this is a well known problem that arises in any certainty equivalence control scheme based on the estimated plant and is really not a drawback associated with the IMC design methodology. Although from time to time, a lot of possible solutions to this problem have been proposed in the adaptive literature, it would be fair to say that, by and large, no satisfactory solution is currently available.

The following Theorem describes the stability and performance properties of the adaptive IMC schemes presented in this chapter. The proof is rather long and technical and is, therefore, relegated to Section 5.5.

Theorem 5.3.1. *Consider the plant (5.1) subject to the adaptive IMC control law (5.4)-(5.7), (5.8), where (5.8) corresponds to any one of the adaptive IMC schemes considered in this chapter and $r(t)$ is a bounded external signal. Then all the signals in the closed loop system are uniformly bounded and the error $y - \hat{y} \to 0$ as $t \to \infty$.*

5.4 Adaptive IMC Design Examples

In this section, we present some examples to illustrate the steps involved in the design of the proposed certainty equivalence adaptive controllers.

Example 5.4.1. (Adaptive Partial Pole Placement) We first consider the plant (5.1) with $Z_0(s) = s + 2$, $R_0(s) = s^2 + s + 1$. Choosing $\Lambda(s) = s^2 + 2s + 2$, the plant parametrization becomes

$$y = \theta^{*T}\phi$$

where

$$\theta^* = [1, 1, 1, 2]^T$$

$$\phi = \left[\frac{s}{s^2 + 2s + 2}[y], \frac{1}{s^2 + 2s + 2}[y], \frac{s}{s^2 + 2s + 2}[u], \frac{1}{s^2 + 2s + 2}[u]\right]^T$$

Choosing $\mathcal{C}_\theta = [-5.0,\ 5.0] \times [-4.0,\ 4.0] \times [0.1,\ 6.0] \times [-6.0,\ 6.0]$, $\gamma = 1$, $Q(s) = \frac{1}{s+4}$ and implementing the adaptive partial pole placement control scheme (5.4)-(5.7), (5.8), with $\theta(0) = [-1.0,\ 2.0,\ 3.0,\ 1.0]^T$ and all other initial conditions set to zero, we obtained the plots shown in Fig. 5.2 for $r(t) = 1.0$ and $r(t) = sin(0.2t)$. From these plots, it is clear that $y(t)$ tracks $\frac{s+2}{(s^2+s+1)(s+4)}[r]$ quite well.

Example 5.4.2. (Adaptive Model Reference Control) Let us now consider the design of an adaptive model reference control scheme for the same plant used in the last example where now the reference model is given by $W_m(s) = \frac{1}{s^2+2s+1}$. The adaptive law (5.4)-(5.7) must now guarantee that the estimated plant is pointwise minimum phase, to ensure which, we now choose the set $\mathcal{C}_\theta$ as $\mathcal{C}_\theta = [-5.0,\ 5.0] \times [-4.0,\ 4.0] \times [0.1,\ 6.0] \times [0.1,\ 6.0]$. All the other design parameters are kept exactly the same as in the last example except that now (5.8) implements the IMC control law (5.9) with $\Lambda_1(s) = s^3 + 2s^2 + 2s + 2$. This choice of a third order $\Lambda_1(s)$ is necessary since from (5.9), it is clear that $n_d = 3$ here. The resulting simulation plots are shown in Fig. 5.3 for $r(t) = 1.0$ and $r(t) = sin(0.2t)$. From these plots, it is clear that the adaptive IMC scheme does achieve model following.

Example 5.4.3. (Adaptive H_2 Optimal Control) The plant that we have considered in the last two examples is minimum phase which would not lead to an interesting H_2 or H_∞ optimal control problem. Thus, for H_2 and H_∞ optimal control , we consider the plant (5.1) with $Z_0(s) = -s+1$, $R_0(s) = s^2+3s+2$. Choosing $\Lambda(s) = s^2 + 2s + 2$, the plant parametrization becomes

$$y = \theta^{*T}\phi$$

where

$$\theta^* = [-1, 0, -1, 1]^T$$

$$\phi = \left[\frac{s}{s^2 + 2s + 2}[y], \frac{1}{s^2 + 2s + 2}[y], \frac{s}{s^2 + 2s + 2}[u], \frac{1}{s^2 + 2s + 2}[u]\right]^T$$

In order to ensure that $\hat{Z}_0(s,t)$ has no zeros on the imaginary axis and its degree does not drop, the projection set $\mathcal{C}_\theta$ is chosen as $\mathcal{C}_\theta = [-5.0,\ 5.0] \times$

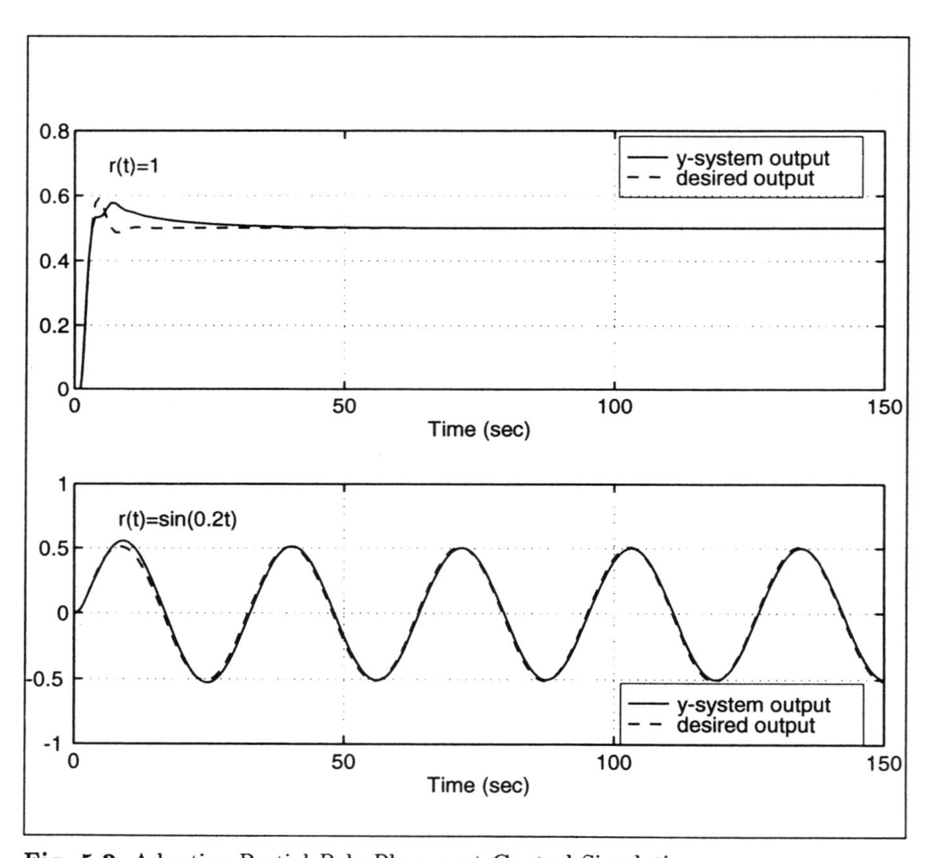

Fig. 5.2. Adaptive Partial Pole Placement Control Simulation

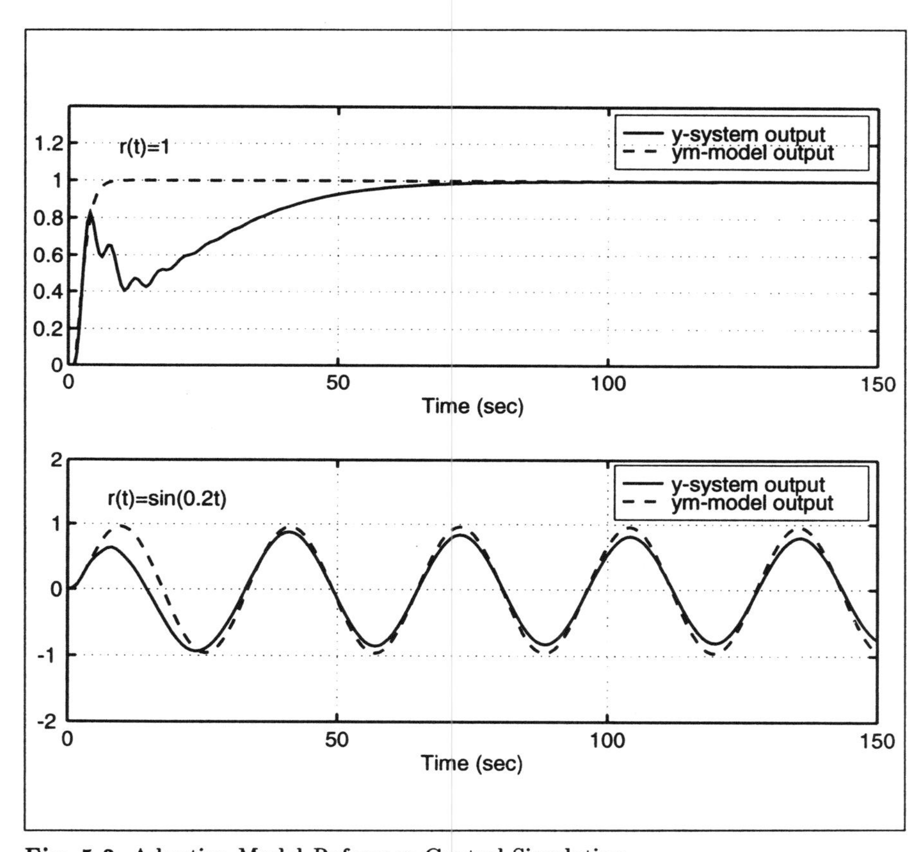

Fig. 5.3. Adaptive Model Reference Control Simulation

$[-4.0,\ 4.0] \times [-6.0,\ -0.1] \times [0.1,\ 6.0]$. Now choosing $\gamma = 1$, $\theta(0) = [-2, 2, -2, 2]^T$, and all other initial conditions equal to zero, we implemented the adaptive law (5.4)-(5.7).

From (5.10), we see that $\hat{Q}(s,t)$ in this case depends on the choice of the command input $r(t)$. Let us choose $r(t)$ to be the unit step function so that $R(s) = R_M(s) = \frac{1}{s}$. Let $\theta(t) = [\theta_1(t), \theta_2(t), \theta_3(t), \theta_4(t)]^T$ denote the estimate of θ^* obtained from the adaptive law (5.4)-(5.7). Then the frozen time estimated plant is given by

$$\hat{P}(s,t) = \frac{\theta_3(t)s + \theta_4(t)}{s^2 + (2 - \theta_1(t))s + (2 - \theta_2(t))}$$

and its right half plane zero $z_p(t)$ is located at $z_p(t) = -\frac{\theta_4(t)}{\theta_3(t)}$. Furthermore

$$\hat{P}(s,t) = \frac{-s + z_p}{s + \bar{z}_p} . \frac{-\theta_3(s + \bar{z}_p)}{s^2 + (2 - \theta_1)s + (2 - \theta_2)}$$

so that $\hat{B}_P(s,t) = \frac{-s+z_p}{s+\bar{z}_p}$ and $\hat{P}_M(s,t) = \frac{-\theta_3(s+\bar{z}_p)}{s^2+(2-\theta_1)s+(2-\theta_2)}$. Hence, from (5.10), we obtain

$$\begin{aligned}\hat{Q}(s,t) &= \frac{s^2 + (2-\theta_1)s + (2-\theta_2)}{-\theta_3(s+\bar{z}_p)} .s \left[\frac{s+\bar{z}_p}{-s+z_p} . \frac{1}{s}\right]_* F(s) \\ &= \frac{s^2 + (2-\theta_1)s + (2-\theta_2)}{-\theta_3(s+\bar{z}_p)} F(s)\end{aligned}$$

It is now clear that to make $\hat{Q}(s,t)$ proper, $F(s)$ must be of relative degree 1. So, let us choose $F(s) = \frac{1}{s+1}$ which results in $n_d = 2$. We now choose $\Lambda_1(s) = s^2 + 2s + 2$, which is of second order, and implement the control law (5.8). The resulting plot is shown in Fig. 5.4. From Fig. 5.4, it is clear that $y(t)$ asymptotically tracks $r(t)$ quite well.

Example 5.4.4. (Adaptive H_∞ Optimal Control) We now design an adaptive H_∞ optimal controller for the same plant considered in the last example. The adaptive law is exactly the same as the one designed earlier and ensures that the frozen-time estimated plant has one and only one right half plane zero. Recall from the previous example that the frozen time estimated plant is given by

$$\hat{P}(s,t) = \frac{\theta_3 s + \theta_4}{s^2 + (2-\theta_1)s + (2-\theta_2)}$$

and the right half plane zero is given by $\hat{b}_1 = -\frac{\theta_4}{\theta_3}$.

Choosing $W(s) = \frac{0.01}{s+0.01}$, from (5.11), we have

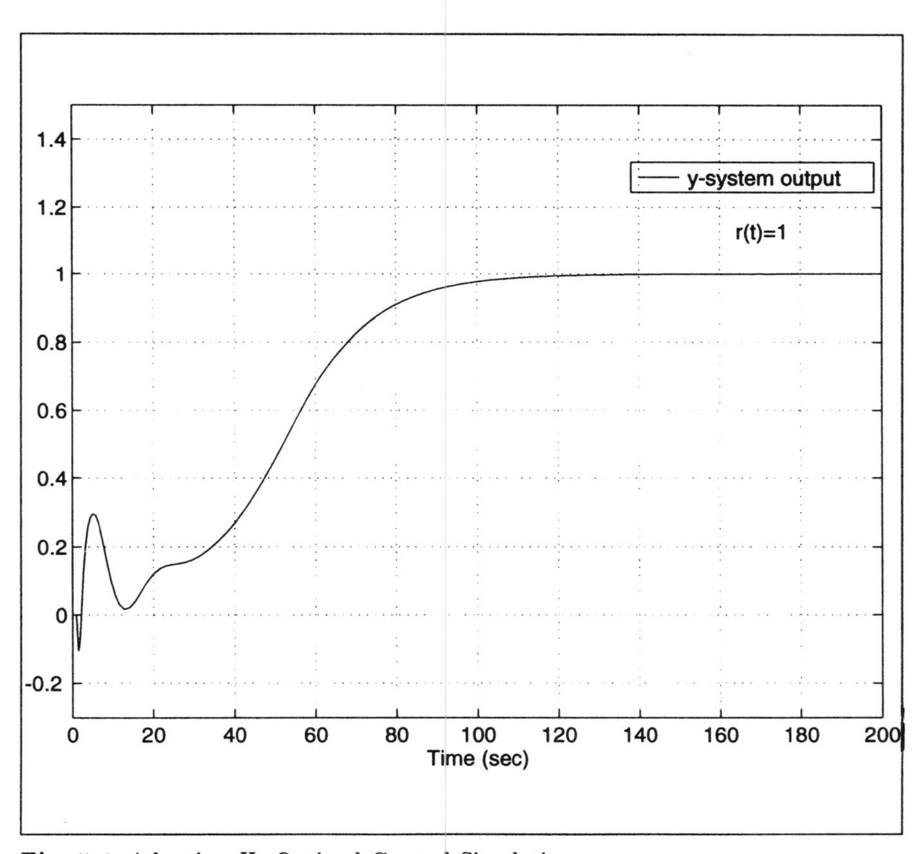

Fig. 5.4. Adaptive H_2 Optimal Control Simulation

$$\begin{aligned}
\hat{Q}(s,t) &= \left[1 - \frac{s+0.01}{\hat{b}_1+0.01}\right] \frac{s^2+(2-\theta_1)s+(2-\theta_2)}{\theta_3 s+\theta_4} F(s) \\
&= \frac{\hat{b}_1 - s}{\theta_3 s+\theta_4} \frac{(s^2+(2-\theta_1)s+(2-\theta_2))}{\hat{b}_1+0.01} F(s) \\
&= -\frac{1}{\theta_3} \frac{(s^2+(2-\theta_1)s+(2-\theta_2))}{\hat{b}_1+0.01} F(s) \text{ (using } \hat{b}_1 = -\tfrac{\theta_4}{\theta_3})
\end{aligned}$$

It is now clear that to make $\hat{Q}(s,t)$ proper, $F(s)$ must be of relative degree 2. So, let us choose $F(s) = \frac{1}{(0.15s+1)^2}$ which results in $n_d = 2$. We now choose $\Lambda_1(s) = s^2+2s+2$ and implement the control law (5.8). Choosing $r(t) = 1.0$ and $r(t) = 0.8sin(0.2t)$, we obtained the plots shown in Fig. 5.5. From these plots, we see that the adaptive H_∞-optimal controller does produce reasonably good tracking.

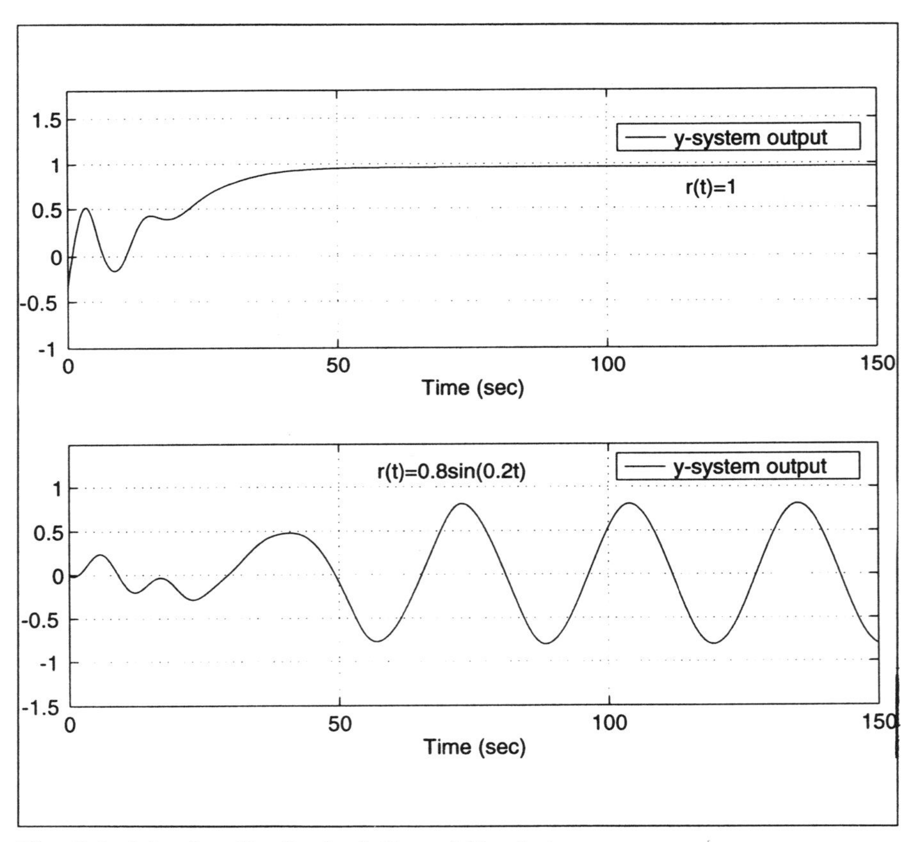

Fig. 5.5. Adaptive H_∞ Optimal Control Simulation

5.5 Stability Proofs of Adaptive IMC Schemes

In this section, we provide the detailed proof of Theorem 5.3.1. The proof makes uses of two technical lemmas. For clarity of presentation, we first state and prove these two lemmas.

Lemma 5.5.1. *In each of the adaptive IMC schemes presented in Section 5.3, the degree of $\hat{Q}_d(s,t)$ in Step 3 of the Certainty Equivalence Design can be made time invariant. Furthermore, for the adaptive H_2 and H_∞ designs, this can be done using a single fixed $F(s)$.*

Proof. The proof of this lemma is relatively straight forward except in the case of adaptive H_2 optimal control. Accordingly, we first discuss the simpler cases before giving a detailed treatment of the more involved one.

For adaptive partial pole placement, the time invariance of the degree of $\hat{Q}_d(s,t)$ follows trivially from the fact that the IMC parameter in this case is time invariant. For model reference adaptive control, the fact that the leading coefficient of $\hat{Z}_o(s,t)$ is not allowed to pass through zero guarantees that the degree of $\hat{Q}_d(s,t)$ is time invariant. Finally, for adaptive H_∞ optimal control, the result follows from the fact that the leading coefficient of $\hat{Z}_o(s,t)$ is not allowed to pass through zero and consequently, $F(s)$ of fixed degree can be chosen to make $\hat{Q}(s,t)$ proper.

We now present the detailed proof for the case of adaptive H_2 optimal control. Let n_r, m_r be the degrees of the denominator and numerator polynomials respectively of $R(s)$. Then, in the expression for $\hat{Q}(s,t)$ in (5.10), it is clear that $\hat{P}_M^{-1}(s,t) = \frac{n\text{th order polynomial}}{l\text{th order polynomial}}$ while $R_M^{-1}(s) = \frac{n_r\text{th order polynomial}}{m_r\text{th order polynomial}}$. Also $\left[\hat{B}_P^{-1}(s,t)R_M(s)\right]_* = \frac{(\bar{n}-1)\text{th order polynomial}}{\bar{n}\text{th order polynomial}}$ where $\bar{n} \leq n_r$, strict inequality being attained when some of the poles of $R_M(s)$ coincide with some of the stable zeros of $\hat{B}_P^{-1}(s,t)$. Moreover, in any case, the $\bar{n}$th order denominator polynomial of $[\hat{B}_P^{-1}(s,t)R_M(s)]_*$ is a factor of the n_rth order numerator polynomial of $R_M^{-1}(s)$. Thus for the $\hat{Q}(s,t)$ given in (5.10), if we disregard $F(s)$, then the degree of the numerator polynomial is $n+n_r-1$ while that of the denominator polynomial is $l+m_r \leq n+n_r-1$. Hence, the degree of $\hat{Q}_d(s,t)$ in Step 3 of the Certainty Equivalence Design can be kept fixed at $(n+n_r-1)$, and this can be achieved with *a single fixed* $F(s)$ of relative degree $n-l+n_r-m_r-1$, provided that the leading coefficient of $\hat{Z}_o(s,t)$ is appropriately constrained.

Remark 5.5.1. Lemma 5.5.1 tells us that the degree of each of the certainty equivalence controllers presented in Section 5.3 can be made time invariant. This is important because, as we will see, it makes it possible to carry out the analysis using standard state-space results on slowly time-varying systems.

Lemma 5.5.2. *At any fixed time t, the coefficients of $\hat{Q}_d(s,t)$, $\hat{Q}_n(s,t)$, and hence the vectors $q_d(t)$, $q_n(t)$ are continuous functions of the estimate $\theta(t)$.*

Proof. Once again, the proof of this lemma is relatively straight forward except in the case of adaptive H_2 optimal control. Accordingly, we first discuss the simpler cases before giving a detailed treatment of the more involved one.

For the case of adaptive partial pole placement control, the continuity follows trivially from the fact that the IMC parameter is independent of $\theta(t)$. For model reference adaptive control, the continuity is immediate from (5.9) and the fact that the leading coefficient of $\hat{Z}_o(s,t)$ is not allowed to pass through zero. Finally for adaptive H_∞ optimal control, we note that the right half plane zero $\hat{b}_1$ of $\hat{Z}_0(s,t)$ is a continuous function of $\theta(t)$. This is a consequence of the fact that the degree of $\hat{Z}_0(s,t)$ cannot drop since its leading coefficient is not allowed to pass through zero. The desired continuity now follows from (5.11).

We now present the detailed proof for the H_2 optimal control case. Since the leading coefficient of $\hat{Z}_o(s,t)$ has been constrained so as not to pass through zero then, for any fixed t, the roots of $\hat{Z}_o(s,t)$ are continuous functions of $\theta(t)$. Hence, it follows that the coefficients of the numerator and denominator polynomials of $[\hat{P}_M(s,t)]^{-1} = [\hat{B}_P(s,t)][\hat{P}(s,t)]^{-1}$ are continuous functions of $\theta(t)$. Moreover, $\left[[\hat{B}_P(s,t)]^{-1}R_M(s)\right]_*$ is the sum of the residues of $[\hat{B}_P(s,t)]^{-1}R_M(s)$ at the poles of $R_M(s)$, which clearly depends continuously on $\theta(t)$ (through the factor $[\hat{B}_P(s,t)]^{-1}$). Since $F(s)$ is fixed and independent of θ, it follows from (5.10) that the coefficients of $\hat{Q}_d(s,t)$, $\hat{Q}_n(s,t)$ depend continuously on $\theta(t)$.

Remark 5.5.2. Lemma 5.5.2 is important because it allows one to translate slow variation (in the L_2 sense) of the estimated parameter vector $\theta(t)$ to slow variation of the controller parameters. Since the stability and robustness proofs of most adaptive schemes rely on results from the stability of slowly time-varying systems, establishing continuity of the controller parameters as a function of the estimated plant parameters (which are known to vary slowly in the L_2 sense) is a crucial ingredient of the analysis.

Proof of Theorem 5.3.1:

The proof of Theorem 5.3.1 is obtained by combining the properties of the adaptive law (5.4)-(5.7) with the properties of the IMC based controller structure. The properties of the adaptive law have already been established in Chapter 4. Indeed, since $\frac{\phi}{m} \in L_\infty$, from Theorem 4.4.1, we see that the adaptive law (5.4)-(5.7) guarantees that (i) $\theta \in L_\infty$ and (ii) $\epsilon, \epsilon n_s, \dot{\theta} \in L_2 \cap L_\infty$ To complete the stability proof, we now turn to the properties of the IMC based controller structure.

The certainty equivalence control law (5.8) can be rewritten as

$$\frac{s^{n_d}}{\Lambda_1(s)}[u] + \beta_1(t)\frac{s^{n_d-1}}{\Lambda_1(s)}[u] + \cdots + \beta_{n_d}(t)\frac{1}{\Lambda_1(s)}[u] = q_n^T(t)\frac{a_{n_d}(s)}{\Lambda_1(s)}[r - \epsilon m^2]$$

where $\beta_1(t), \beta_2(t), \cdots, \beta_{n_d}(t)$ are the time-varying coefficients of $\hat{Q}_d(s,t)$. Defining $x_1 = \frac{1}{\Lambda_1(s)}[u], x_2 = \frac{s}{\Lambda_1(s)}[u], \cdots, x_{n_d} = \frac{s^{n_d-1}}{\Lambda_1(s)}[u], X \triangleq [x_1, x_2, \cdots, x_{n_d}]^T$, the above equation can be rewritten as

$$\dot{X} = A(t)X + Bq_n^T(t)\frac{a_{n_d}(s)}{\Lambda_1(s)}[r - \epsilon m^2] \tag{5.12}$$

where

$$A(t) \triangleq \begin{bmatrix} 0 & 1 & 0 & \cdot & \cdot & 0 \\ 0 & 0 & 1 & 0 & \cdot & 0 \\ \cdot & \cdot & \cdot & \cdot & \cdot & \cdot \\ \cdot & \cdot & \cdot & \cdot & \cdot & \cdot \\ -\beta_{n_d}(t) & -\beta_{n_d-1}(t) & \cdot & \cdot & \cdot & -\beta_1(t) \end{bmatrix}$$

$$B \triangleq \begin{bmatrix} 0 \\ 0 \\ \cdot \\ 0 \\ 1 \end{bmatrix}$$

Since the time-varying polynomial $\hat{Q}_d(s,t)$ is pointwise Hurwitz, it follows that for any *fixed* t, the eigenvalues of $A(t)$ are in the open left half plane. Moreover, since the coefficients of $\hat{Q}_d(s,t)$ are continuous functions of $\theta(t)$ (Lemma 5.5.2) and $\theta(t) \in \mathcal{C}_\theta$, a compact set, it follows that $\exists\ \sigma_s > 0$ such that

$$\mathrm{Re}\{\lambda_i(A(t))\} \leq -\sigma_s\ \forall\ t \geq 0 \text{ and } i = 1, 2, \cdots, n_d.$$

The continuity of the elements of $A(t)$ with respect to $\theta(t)$ and the fact that $\dot{\theta} \in L_2$ together imply that $\dot{A}(t) \in L_2$. Hence, it follows from Theorem 2.4.9(i) that the equilibrium state $x_e = 0$ of $\dot{x} = A(t)x$ is exponentially stable, i.e. there exist $c_o, p_o > 0$ such that the state transition matrix $\Phi(t,\tau)$ corresponding to the homogeneous part of (5.12) satisfies

$$||\Phi(t,\tau)|| \leq c_o e^{-p_o(t-\tau)}\ \forall\ t \geq \tau \tag{5.13}$$

From the identity $u = \frac{\Lambda_1(s)}{\Lambda_1(s)}[u]$, it is easy to see that the control input u can be rewritten as

$$u = v^T(t)X + q_n^T(t)\frac{a_{n_d}(s)}{\Lambda_1(s)}[r - \epsilon m^2] \tag{5.14}$$

$$\begin{aligned} \text{where } v(t) &= [\lambda_{n_d} - \beta_{n_d}(t), \lambda_{n_d-1} - \beta_{n_d-1}(t), \cdots, \lambda_1 - \beta_1(t)]^T \\ \text{and } \Lambda_1(s) &= s^{n_d} + \lambda_1 s^{n_d-1} + \cdots + \lambda_{n_d}. \end{aligned}$$

Also, using (5.14) in the plant equation (5.1), we obtain

$$y = \frac{Z_0(s)}{R_0(s)}\left[v^T(t)X + q_n^T(t)\frac{a_{n_d}(s)}{\Lambda_1(s)}[r - \epsilon m^2]\right] \tag{5.15}$$

Now let $\delta \in (0, p_o)$ be chosen such that $\frac{1}{R_0(s)}, \frac{1}{\Lambda_1(s)}, \frac{1}{\Lambda(s)}$ are analytic in $\mathcal{R}e[s] \geq -\frac{\delta}{2}$, and define the fictitious signal $m_f(t)$ by

$$m_f(t) = 1.0 + ||u_t||_2^\delta + ||y_t||_2^\delta \tag{5.16}$$

We now take truncated exponentially weighted norms on both sides of (5.14),(5.15) and make use of Lemma 2.3.4 and Lemma 2.3.3(i), while observing that $v(t)$,$q_n(t)$,$r(t) \in L_\infty$, to obtain[3]

$$||u_t||_2^\delta \leq c + c||(\varepsilon m^2)_t||_2^\delta \tag{5.17}$$

$$||y_t||_2^\delta \leq c + c||(\varepsilon m^2)_t||_2^\delta \tag{5.18}$$

which together with (5.16) imply that

$$m_f(t) \leq c + c||(\varepsilon m^2)_t||_2^\delta \tag{5.19}$$

Now by (5.3) and Lemma 2.3.3(ii), we have

$$m = \sqrt{1 + \phi^T \phi} \leq c m_f.$$

Hence, squaring both sides of (5.19) we obtain

$$\begin{aligned} m_f^2(t) &\leq c + c \int_0^t e^{-\delta(t-\tau)} \varepsilon^2 m^2 m_f^2(\tau) d\tau \\ \Rightarrow m_f^2(t) &\leq c + c \int_0^t e^{-\delta(t-s)} \varepsilon^2(s) m^2(s) \left(e^{c \int_s^t \varepsilon^2 m^2 d\tau} \right) ds \end{aligned}$$

(using Lemma 2.3.5, i.e. the Bellman-Gronwall Lemma)

Since $\varepsilon m \in L_2$, it follows using Theorem 2.3.3(iii) that $m_f \in L_\infty$, which in turn implies that $m \in L_\infty$. Since $\frac{\phi}{m}$ is bounded, it follows that $\phi \in L_\infty$. Thus $\varepsilon m^2 = -\tilde{\theta}^T \phi$ is also bounded so that from (5.12), we obtain $X \in L_\infty$. From (5.14), (5.15), we can now conclude that u, $y \in L_\infty$. This establishes the boundedness of all the closed loop signals in the adaptive IMC Scheme. Since $y - \hat{y} = \varepsilon m^2$ and $\varepsilon m \in L_2 \cap L_\infty$, $m \in L_\infty$, it follows that $y - \hat{y} \in L_2 \cap L_\infty$. Now $\frac{d}{dt}(y - \hat{y}) = \dot{y} - \frac{d}{dt}\left(\theta^T \phi\right) = \dot{y} - \dot{\theta}^T \phi - \theta^T \dot{\phi}$. Since $y, u \in L_\infty$, it follows from (5.1), (5.3) that $\dot{y}, \dot{\phi} \in L_\infty$. Thus $\frac{d}{dt}(y - \hat{y})$ is bounded and from Corollary 2.2.1, we obtain $y - \hat{y} \to 0$ as $t \to \infty$. This completes the proof.

[3] In the rest of this proof, 'c' is the generic symbol for a positive constant. The exact value of such a constant can be determined (for a quantitative robustness result) as in [42, 15]. However, for the qualitative presentation here, the exact values of these constants are not important.

CHAPTER 6
ROBUST PARAMETER ESTIMATION

6.1 Introduction

In the last two chapters, we have designed and analyzed on-line parameter estimators and adaptive control schemes under the assumption that there are no modelling errors. Such an assumption is unrealistic since in the real world, modelling errors such as disturbances, sensor noise, unmodelled dynamics, nonlinearities, etc. will most likely be present. The aim of this chapter is to first examine how the on-line parameter estimators of Chapter 4 behave in the presence of modelling errors. Thereafter we will explore several approaches for correcting possible unsatisfactory behaviour. Parameter estimators that can behave satisfactorily even in the presence of modelling errors are called *Robust Parameter Estimators*. Thus this chapter is primarily concerned with the design and analysis of ***robust*** parameter estimators.

We begin in Section 6.2 by presenting some simple examples to show that the parameter estimators of Chapter 4 can exhibit instability in the presence of small modelling errors. This motivates the introduction of robustifying modifications in the parameter estimators. These modifications are first illustrated on a simple scalar example in Section 6.3, following which the full-blown design of robust parameter estimators for a general plant is taken up in Section 6.4.

6.2 Instability Phenomena in Adaptive Parameter Estimation

In this section, we present simple examples to show that the parameter estimators designed in Chapter 4 can become unstable in the presence of modelling errors. Let us consider the problem of estimating the unknown parameter θ^* from measurements of the signals $u(t)$ and $y(t)$ where $u(t), y(t)$ are related by

$$y(t) = \theta^* u(t) + d(t) \tag{6.1}$$

and $d(t)$ is an unknown bounded disturbance. If $d(t) \equiv 0$, then we obtain the model

$$y(t) = \theta^* u(t) \tag{6.2}$$

based on which *normalized* and *unnormalized* adaptive laws have been designed in Chapter 4 for estimating θ^*. Our objective here is to examine how these adaptive laws behave when $d \neq 0$. We consider the unnormalized and normalized cases separately.

Example 6.2.1. (Unnormalized Adaptive Law) When $d \equiv 0$ in (6.1), then from (4.6) in Chapter 4, the unnormalized gradient adaptive law is given by

$$\dot{\theta} = \gamma \epsilon_1 u, \; \epsilon_1 = y - \theta u, \; \gamma > 0 \tag{6.3}$$

Furthermore, as shown in Section 4.2.1, if $d \equiv 0$ and $u, \dot{u} \in L_\infty$, then $\theta, \epsilon_1 \in L_\infty$ and $\epsilon_1(t) \to 0$ as $t \to \infty$. Let us now analyze the same adaptive law when $d \neq 0$. Then, we have

$$\epsilon_1 = \theta^* u + d - \theta u = -\tilde{\theta} u + d, \; \tilde{\theta} \triangleq \theta - \theta^* \tag{6.4}$$

Substituting (6.4) into (6.3), we obtain

$$\dot{\tilde{\theta}} = -\gamma u^2 \tilde{\theta} + \gamma d u \tag{6.5}$$

Now, as in Chapter 4, we consider the Lyapunov-like function $V(\tilde{\theta}) = \frac{\tilde{\theta}^2}{2\gamma}$ whose time derivative along the solution of (6.5) is given by

$$\begin{aligned} \dot{V} &= -u^2\tilde{\theta}^2 + d\tilde{\theta}u \\ &= -\frac{\tilde{\theta}^2 u^2}{2} - \frac{1}{2}(\tilde{\theta}u - d)^2 + \frac{d^2}{2}. \end{aligned}$$

Since the above expression for $\dot{V}$ is sign indefinite, we cannot conclude anything about the boundedness of $\tilde{\theta}$. Since Lyapunov-type analysis provides only *sufficient* conditions for boundedness, one may argue that the boundedness of $\tilde{\theta}$ could perhaps be established by using another Lyapunov-like function. However, by explicitly solving (6.5), we can show that this is not going to be the case:

Indeed, choosing $\theta^* = 2$, $\gamma = 1$, $u(t) = (1+t)^{-\frac{1}{2}}$, $\theta(0) = 1$ and $d(t) = (1+t)^{-\frac{1}{4}}(\frac{5}{4} - 2(1+t)^{-\frac{1}{4}})$, we can solve (6.5) to obtain

$$\theta(t) = (1+t)^{\frac{1}{4}}. \tag{6.6}$$

This shows that although $u, d \in L_\infty$ and $d(t)$, in fact, vanishes asymptotically with time nevertheless, $\theta(t)$ and hence $\tilde{\theta}(t)$ drift to infinity as $t \to \infty$ so that no analysis, Lyapunov or otherwise, could possibly be used to establish the boundedness of $\tilde{\theta}$. Thus the boundedness properties of $\theta, \tilde{\theta}$ established in Chapter 4 for the adaptive law (6.3) are destroyed in the presence of modelling errors. The instability phenomenon observed in this example is referred to in the adaptive control literature as *parameter drift* [19].

Example 6.2.2. (Normalized Adaptive Law) When $d \equiv 0$ in (6.1), then from (4.40) in Chapter 4, the normalized gradient adaptive law is given by

$$\dot{\theta} = \gamma \epsilon u, \; \epsilon = \frac{y - \theta u}{m^2}, \; m^2 = 1 + u^2, \; \gamma > 0 \tag{6.7}$$

Furthermore, as shown in Section 4.4.1, if $d \equiv 0$, then $\theta \in L_\infty$, and $\epsilon m, \dot{\theta} \in L_2 \cap L_\infty$. Let us now consider the same adaptive law when $d \neq 0$. Then, we have

$$\epsilon = \frac{\theta^* u + d - \theta u}{1 + u^2} = -\tilde{\theta}\frac{u}{1 + u^2} + \frac{d}{1 + u^2} \tag{6.8}$$

Substituting (6.8) into (6.7), we obtain

$$\dot{\tilde{\theta}} = -\gamma \frac{u^2}{1 + u^2}\tilde{\theta} + \gamma \frac{du}{1 + u^2} \tag{6.9}$$

Now, choosing $\theta^* = 2$, $\gamma = 1$, $u(t) = (1 + t)^{-\frac{1}{2}}$, $\theta(0) = \frac{1}{2}$ and $d(t) = \frac{5}{4}(1 + t)^{-\frac{1}{4}} - 2(1 + t)^{-\frac{1}{2}}$, we can solve (6.9) explicitly to obtain

$$\theta(t) = \frac{(1 + t)^{\frac{5}{4}}}{(2 + t)}$$

which shows that the boundedness properties of $\theta, \tilde{\theta}$ established in Chapter 4 for the adaptive law (6.7) are destroyed in the presence of modelling errors.

6.3 Modifications for Robustness: Simple Examples

Instability examples such as the ones above attracted a considerable amount of interest in the adaptive literature during the late seventies and early eighties [11, 16, 17, 37]. During this period, it was realized that adaptive schemes designed with the ideal plant in mind could exhibit instability in the presence of modelling errors. Furthermore, the cause of instability was correctly identified as the adaptive law whose presence made the closed loop system nonlinear and, therefore, more susceptible to the effect of modelling errors. This led to a period of intense research activity aimed at modifying the adaptive laws so as to enable them to guarantee certain desirable properties even in the presence of modelling errors (see [19] and the references therein). Such adaptive laws are referred to as *Robust Adaptive Laws.*

Although most of the modifications for robustness were originally adhoc in nature and bore very little semblance to each other, research during the last decade has shown that it is possible to design and analyze most of them using a unified theory [18, 15, 19]. Our discussion of robust adaptive laws in this chapter will be based on this unified theory. In order to gain a thorough understanding of the underlying concepts involved, in this section, we will first design and analyze robust adaptive laws only for a simple scalar example.

Therafter, in the next section, we will take up the design of robust adaptive laws for estimating the parameters of a general nth order plant.

Once again, let us consider the problem of estimating the unknown parameter θ^* from measurements of the signals $u(t)$ and $y(t)$ where $u(t)$, $y(t)$ are related by

$$y(t) = \theta^* u(t) + \eta(t) \tag{6.10}$$

and η is an unknown modelling error term. When $\eta = 0$ and $u \in L_\infty$, we can use the adaptive law (6.3) to guarantee the following two properties: (i) $\epsilon_1, \theta, \dot{\theta} \in L_\infty$ and (ii) $\epsilon_1, \dot{\theta} \in L_2$. If, in addition, $\dot{u} \in L_\infty$, the adaptive law also guarantees that $\epsilon_1, \dot{\theta} \to 0$ as $t \to \infty$. When $\eta = 0$ but $u \notin L_\infty$, we use the normalized adaptive law (6.7) to guarantee the following two properties: (i) $\epsilon, \epsilon m, \theta, \dot{\theta} \in L_\infty$ (ii) $\epsilon, \epsilon m, \dot{\theta} \in L_2$. The properties (i) and (ii) established for $\eta \equiv 0$ are referred to as the *ideal properties* of the adaptive laws. When $\eta \neq 0$, these properties do not hold as evident from the examples presented in Section 6.2. Our objective here is, therefore, to modify the adaptive laws so that the properties of the modified adaptive laws are as close as possible to the ideal properties (i) and (ii).

We first consider the simpler case where η is due to a bounded disturbance and $u \in L_\infty$ and explore different approaches for modifying the adaptive law (6.3). To this end, let $d_0 > 0$ be an upper bound for $|\eta|$, i.e. $d_0 \geq \sup_t |\eta(t)|$.

6.3.1 Leakage

The principal idea behind leakage is to modify the adaptive law so that $\dot{V}$ becomes negative in the space of the parameter estimates when these parameter estimates exceed certain bounds. Instead of (6.3), the adaptive law with leakage becomes

$$\dot{\theta} = \gamma\epsilon_1 u - \gamma w\theta,\ \epsilon_1 = y - \theta u \tag{6.11}$$

where $w(t) \geq 0$ is a design function which converts the "pure" integration in (6.3) to a "leaky" integration as in (6.11). This is the reason why this modification is referred to as *leakage*. The design variable $w(t) \geq 0$ is to be chosen so that for $V \geq V_0 > 0$ and some V_0 which may depend on the bound for $d(t)$, $\dot{V} \leq 0$ so that Theorem 2.4.3 on uniform boundedness and uniform ultimate boundedness can be invoked. Let us now look at some specific choices of $w(t)$:

(a) **σ-modification** [16]: In this modification, $w(t)$ is chosen to be a positive constant i.e. $w(t) = \sigma > 0\ \forall\, t \geq 0$. With this choice, the adaptive law (6.11) becomes

$$\dot{\theta} = \gamma\epsilon_1 u - \gamma\sigma\theta,\ \epsilon_1 = y - \theta u \tag{6.12}$$

Noting that $\dot{\theta} = \frac{d}{dt}[\tilde{\theta} + \theta^*] = \dot{\tilde{\theta}}$ and substituting for y from (6.10), the above adaptive law can be rewritten as

$$\dot{\tilde{\theta}} = \gamma\epsilon_1 u - \gamma\sigma\theta,\ \epsilon_1 = -\tilde{\theta}u + \eta \tag{6.13}$$

Now consider the Lyapunov-like function $V = \frac{\tilde{\theta}^2}{2\gamma}$ whose derivative along the solution of (6.13) is given by

$$\begin{aligned} \dot{V} &= \epsilon_1 \tilde{\theta} u - \sigma \tilde{\theta} \theta \\ &= \epsilon_1(-\epsilon_1 + \eta) - \sigma \tilde{\theta} \theta \text{ (using the expression for } \epsilon_1 \text{ from (6.13))} \\ &= -\epsilon_1^2 + \epsilon_1 \eta - \sigma \tilde{\theta} \theta \\ &\leq -\epsilon_1^2 + |\epsilon_1| d_0 - \sigma \tilde{\theta} \theta \end{aligned} \tag{6.14}$$

Now by a simple completion of squares, we obtain

$$\begin{aligned} -\epsilon_1^2 + |\epsilon_1| d_0 &= -\frac{\epsilon_1^2}{2} - \frac{1}{2}[\epsilon_1^2 - 2|\epsilon_1| d_0] \\ &= -\frac{\epsilon_1^2}{2} - \frac{1}{2}[|\epsilon_1| - d_0]^2 + \frac{d_0^2}{2} \\ &\leq -\frac{\epsilon_1^2}{2} + \frac{d_0^2}{2} \end{aligned} \tag{6.15}$$

$$\begin{aligned} \text{Also } -\sigma \tilde{\theta} \theta &= -\sigma \tilde{\theta}(\tilde{\theta} + \theta^*) \\ &\leq -\sigma \tilde{\theta}^2 + \sigma |\tilde{\theta}||\theta^*| \\ &= -\frac{\sigma}{2}\tilde{\theta}^2 - \frac{\sigma}{2}(\tilde{\theta}^2 - 2|\tilde{\theta}||\theta^*|) \\ &= -\frac{\sigma}{2}\tilde{\theta}^2 - \frac{\sigma}{2}(|\tilde{\theta}| - |\theta^*|)^2 + \frac{\sigma}{2}|\theta^*|^2 \\ &\leq -\frac{\sigma}{2}\tilde{\theta}^2 + \frac{\sigma}{2}|\theta^*|^2 \end{aligned} \tag{6.16}$$

Thus $\dot{V} \leq -\frac{\epsilon_1^2}{2} - \frac{\sigma}{2}\tilde{\theta}^2 + \frac{d_0^2}{2} + \frac{\sigma}{2}|\theta^*|^2$. Adding and subtracting the term αV where $\alpha > 0$ is arbitrary, we obtain

$$\dot{V} \leq -\alpha V - \frac{\epsilon_1^2}{2} - (\sigma - \frac{\alpha}{\gamma})\frac{\tilde{\theta}^2}{2} + \frac{d_0^2}{2} + \frac{\sigma}{2}|\theta^*|^2.$$

Choosing $0 < \alpha \leq \sigma\gamma$, we have

$$\dot{V} \leq -\alpha V + \frac{d_0^2}{2} + \frac{\sigma}{2}|\theta^*|^2 \tag{6.17}$$

which implies that for $V \geq V_0 = \frac{1}{2\alpha}(d_0^2 + \sigma|\theta^*|^2)$, $\dot{V} \leq 0$. Therefore, using Theorem 2.4.3, we have $\tilde{\theta}, \theta \in L_\infty$ which together with $u \in L_\infty$ imply that $\epsilon_1, \dot{\theta} \in L_\infty$. Thus using the σ-modification, we managed to extend property (i) for the ideal case to the case where $\eta \neq 0$ provided $\eta \in L_\infty$. In addition, by integrating (6.17), we can establish that

$$V(\tilde{\theta}(t)) = \frac{\tilde{\theta}^2(t)}{2\gamma} \leq e^{-\alpha t}\frac{\tilde{\theta}^2(0)}{2\gamma} + \frac{1}{2\alpha}(d_0^2 + \sigma|\theta^*|^2)$$

which shows that $\tilde{\theta}(t)$ converges exponentially to the residual set

$$D_\sigma \triangleq \left\{\tilde{\theta} \in R \mid \tilde{\theta}^2 \leq \frac{\gamma}{\alpha}(d_0^2 + \sigma|\theta^*|^2)\right\}.$$

Let us now examine whether property (ii) for the ideal case namely $\epsilon_1, \dot{\theta} \in L_2$ can be extended to the case where $\eta \neq 0$. Now

$$-\sigma\tilde{\theta}\theta = -\sigma(\theta - \theta^*)\theta \leq -\frac{\sigma}{2}|\theta|^2 + \frac{\sigma}{2}|\theta^*|^2$$

so that from (6.14) and (6.15) we obtain

$$\dot{V} \leq -\frac{\epsilon_1^2}{2} - \frac{\sigma}{2}|\theta|^2 + \frac{d_0^2}{2} + \frac{\sigma}{2}|\theta^*|^2.$$

Integrating both sides of the above inequality, we obtain

$$\begin{aligned}
\int_t^{t+T} \epsilon_1^2(\tau)d\tau + \int_t^{t+T} \sigma\theta^2(\tau)d\tau &\leq (d_0^2 + \sigma|\theta^*|^2)T + 2[V(t) - V(t+T)] \\
&\leq c_0(d_0^2 + \sigma)T + c_1 \\
\text{where } c_0 &= \max[1, |\theta^*|^2] \\
c_1 &= 2 \sup_{t,T \geq 0} [V(t) - V(t+T)].
\end{aligned}$$

Thus $\epsilon_1, \sqrt{\sigma}|\theta| \in \mathcal{S}(d_0^2 + \sigma)$. Since $(a+b)^2 \leq 2a^2 + 2b^2$, from (6.12) we obtain

$$|\dot{\theta}|^2 \leq 2\gamma^2\epsilon_1^2 u^2 + 2\gamma^2\sigma^2\theta^2.$$

Since $u, \sigma \in L_\infty$, it follows that $\dot{\theta} \in \mathcal{S}(d_0^2 + \sigma)$. Therefore, the ideal property (ii) extends to $(ii)'$ $\epsilon_1, \dot{\theta} \in \mathcal{S}(d_0^2 + \sigma)$.

Let us recall that the L_2 property of ϵ_1 was used in the ideal case to show that $\epsilon_1(t) \to 0$ as $t \to \infty$. Here, even when $\eta = 0$, we cannot guarantee that $\epsilon_1, \dot{\theta} \in L_2$ or $\epsilon_1(t) \to 0$ as $t \to \infty$ unless $\sigma = 0$. Thus robustness is achieved at the expense of destroying the ideal property given by (ii). This drawback of the σ-modification was what motivated the *switching-σ* modification to be presented next.

(b) **Switching σ-modification** [20]: In this case, we choose

$$w(t) = \sigma_s \tag{6.18}$$

$$\sigma_s = \begin{cases} 0 & \text{if } |\theta| < M_0 \\ \sigma_0 & \text{if } |\theta| \geq M_0 \end{cases} \tag{6.19}$$

where $M_0, \sigma_0 > 0$ are design parameters and $M_0 > |\theta^*|$ i.e. M_0 is a known upper bound for $|\theta^*|$. The above choice for $w(t)$ is very intuitive — when the parameter estimates exceed certain pre-specified bounds, signalling a possible parameter drift, turn on the σ-modification, otherwise let the σ-modification be inactive. The switching in (6.19) is discontinuous and can potentially lead to problems with the existence and uniqueness of solutions to (6.11) [35]. To avoid any such problems, we can make the switching continuous by defining

$$\sigma_s = \begin{cases} 0 & \text{if } |\theta(t)| \leq M_0 \\ \sigma_0 \left(\frac{|\theta(t)|}{M_0} - 1 \right) & \text{if } M_0 < |\theta(t)| \leq 2M_0 \\ \sigma_0 & \text{if } |\theta(t)| > 2M_0 \end{cases} \tag{6.20}$$

The graph of σ_s versus $|\theta|$ is sketched in Fig. 6.1. As in the case of the σ

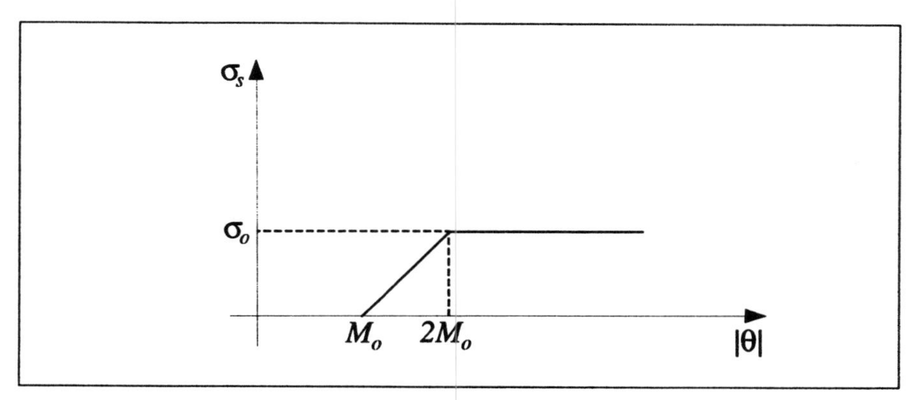

Fig. 6.1. The Plot of σ_s versus $|\theta|$

modification, the adaptive law (6.11), (6.18) can be rewritten as

$$\dot{\tilde{\theta}} = \gamma \epsilon_1 u - \gamma \sigma_s \theta, \; \epsilon_1 = -\tilde{\theta} u + \eta \tag{6.21}$$

Furthermore, starting with $V = \frac{\tilde{\theta}^2}{2\gamma}$, taking its time derivative along solutions of (6.21) and completing squares, we obtain

$$\dot{V} \leq -\frac{\epsilon_1^2}{2} - \sigma_s \tilde{\theta}\theta + \frac{d_0^2}{2} \tag{6.22}$$

$$\begin{aligned} \text{Now } \sigma_s \tilde{\theta}\theta &= \sigma_s(\theta - \theta^*)\theta \\ &\geq \sigma_s|\theta|^2 - \sigma_s|\theta^*||\theta| \\ &= \sigma_s|\theta|(|\theta| - M_0 + M_0 - |\theta^*|) \\ &= \sigma_s|\theta|(|\theta| - M_0) + \sigma_s|\theta|(M_0 - |\theta^*|) \\ &\geq 0 \text{ (by (6.20) and the fact that } M_0 > |\theta^*|) \end{aligned} \tag{6.23}$$

Thus the switching σ can only make $\dot{V}$ more negative. We now show that the switching σ modification can guarantee similar properties as the (fixed) σ modification. Now $-\sigma_s\tilde{\theta}\theta = -\sigma_0\tilde{\theta}\theta + (\sigma_0 - \sigma_s)\tilde{\theta}\theta$. Furthermore, from (6.20)

$$\sigma_0 - \sigma_s = \begin{cases} \sigma_0 & \text{if } |\theta| \leq M_0 \\ \sigma_0 \left(2 - \frac{|\theta|}{M_0}\right) & \text{if } M_0 < |\theta| \leq 2M_0 \\ 0 & \text{if } |\theta| > 2M_0 \end{cases}$$

Since $|\tilde{\theta}| = |\theta - \theta^*| \leq |\theta| + |\theta^*| < |\theta| + M_0$, it follows that

$$(\sigma_0 - \sigma_s)\tilde{\theta}\theta \leq \begin{cases} 2\sigma_0 M_0^2 & \text{if } |\theta| \leq M_0 \\ 6\sigma_0 M_0^2 & \text{if } M_0 < |\theta| \leq 2M_0 \\ 0 & \text{if } |\theta| > 2M_0 \end{cases}$$

Thus $-\sigma_s\tilde{\theta}\theta \leq -\sigma_0\tilde{\theta}\theta + 6\sigma_0 M_0^2$ so that from (6.22)

$$\dot{V} \leq -\frac{\epsilon_1^2}{2} - \sigma_0\tilde{\theta}\theta + 6\sigma_0 M_0^2 + \frac{d_0^2}{2}.$$

From the above inequality, it follows that the boundedness of $\tilde{\theta}, \epsilon_1, \dot{\theta}$ can be obtained using the same procedure as in the case of the fixed σ modification.

Now, integrating (6.22), we obtain

$$2\int_t^{t+T} \sigma_s\tilde{\theta}\theta d\tau + \int_t^{t+T} \epsilon_1^2(\tau)d\tau \leq d_0^2 T + c_1$$

where $c_1 = 2\sup_{t,T\geq 0}[V(t) - V(t+T)]$. Since $\sigma_s\tilde{\theta}\theta \geq 0$, it follows that $\sqrt{\sigma_s\tilde{\theta}\theta}, \epsilon_1 \in \mathcal{S}(d_0^2)$. Now, from (6.23), we have

$$\sigma_s\tilde{\theta}\theta \geq \sigma_s|\theta|(M_0 - |\theta^*|)$$

which implies that

$$\sigma_s^2|\theta|^2 \leq \frac{(\sigma_s\tilde{\theta}\theta)^2}{(M_0 - |\theta^*|)^2} \leq c\sigma_s\tilde{\theta}\theta$$

where $c > 0$ is some constant. Hence it follows that $\sigma_s|\theta| \in \mathcal{S}(d_0^2)$. Furthermore, from (6.11), (6.18), we have

$$|\dot{\theta}|^2 \leq 2\gamma^2\epsilon_1^2 u^2 + 2\gamma^2\sigma_s^2|\theta|^2$$

from which we can conclude that $\dot{\theta} \in \mathcal{S}(d_0^2)$. Hence the adaptive law with switching σ guarantees that $(ii)'$ $\epsilon_1, \dot{\theta} \in \mathcal{S}(d_0^2)$. We note that if $\eta = 0$ then $d_0 = 0$ and the property $(ii)'$ reverts to (ii), i.e. with the switching-σ modification, the L_2 properties are recovered should the modelling errors disappear.

(c) ϵ_1 **modification** [33]: Another possible choice for $w(t)$ aimed at removing the main drawback of the fixed-σ modification is

$$w(t) = |\epsilon_1|\nu_0$$

where $\nu_0 > 0$ is a design constant. With this choice for $w(t)$, the adaptive law (6.11) becomes

$$\dot{\theta} = \gamma\epsilon_1 u - \gamma|\epsilon_1|\nu_0\theta, \quad \epsilon_1 = y - \theta u \tag{6.24}$$

which can be rewritten in terms of the parameter error $\tilde{\theta}$ as

$$\dot{\tilde{\theta}} = \gamma\epsilon_1 u - \gamma|\epsilon_1|\nu_0\theta, \ \epsilon_1 = -\tilde{\theta}u + \eta \tag{6.25}$$

Now consider the Lyapunov-like function $V(\tilde{\theta}) = \frac{\tilde{\theta}^2}{2\gamma}$ and evaluate its derivative along the solution of (6.25) to obtain

$$\begin{aligned}
\dot{V} &= \epsilon_1\tilde{\theta}u - |\epsilon_1|\nu_0\tilde{\theta}\theta \\
&= \epsilon_1[-\epsilon_1 + \eta] - |\epsilon_1|\nu_0\tilde{\theta}\theta \qquad (6.26)\\
&\quad \text{(using the expression for } \epsilon_1 \text{ in (6.25))} \\
&\leq -|\epsilon_1|(|\epsilon_1| + \nu_0\frac{\tilde{\theta}^2}{2} - \nu_0\frac{|\theta^*|^2}{2} - d_0)
\end{aligned}$$

where the above inequality is obtained by using

$$-\nu_0\tilde{\theta}\theta \leq -\nu_0\frac{\tilde{\theta}^2}{2} + \nu_0\frac{\theta^{*2}}{2}.$$

It is now clear that for $\nu_0\frac{\tilde{\theta}^2}{2} \geq \nu_0\frac{|\theta^*|^2}{2} + d_0$, i.e. for $V \geq V_0 \triangleq \frac{1}{\gamma\nu_0}(\nu_0\frac{|\theta^*|^2}{2} + d_0)$, we have $\dot{V} \leq 0$ which implies that V and, therefore, $\theta, \tilde{\theta} \in L_\infty$. Since $\epsilon_1 = -\tilde{\theta}u + \eta$ and $u, \eta \in L_\infty$, it follows that $\epsilon_1 \in L_\infty$ and from (6.24), we conclude that $\dot{\theta} \in L_\infty$. Hence property (i) is guaranteed by the ϵ_1-modification despite the presence of $\eta \neq 0$. Let us now turn to property (ii). Now from (6.26),

$$\begin{aligned}
\dot{V} &= -\epsilon_1^2 + \epsilon_1\eta - |\epsilon_1|\nu_0\tilde{\theta}(\tilde{\theta} + \theta^*) \\
&\leq -\epsilon_1^2 + |\epsilon_1|d_0 - |\epsilon_1|\nu_0\tilde{\theta}^2 - |\epsilon_1|\nu_0\theta^*\tilde{\theta} \\
&\leq -\frac{\epsilon_1^2}{2} + \frac{d_0^2}{2} - |\epsilon_1|\frac{\nu_0}{2}\tilde{\theta}^2 + |\epsilon_1|\frac{\nu_0}{2}|\theta^*|^2 \\
&\quad \text{(completing squares and using } \pm\theta^*\tilde{\theta} \leq \tfrac{\theta^{*2}+\tilde{\theta}^2}{2}\text{)} \\
&\leq -\frac{\epsilon_1^2}{4} - \frac{\epsilon_1^2}{4} + |\epsilon_1|\frac{\nu_0}{2}|\theta^*|^2 + \frac{d_0^2}{2} \\
&\leq -\frac{\epsilon_1^2}{4} + \nu_0^2\frac{\theta^{*4}}{4} + \frac{d_0^2}{2} \ \text{(completing squares)}
\end{aligned}$$

Now integrating on both sides, we establish that $\epsilon_1 \in \mathcal{S}(d_0^2 + \nu_0^2)$. Furthermore, $u, \theta \in L_\infty$ implies that $|\dot{\theta}| \leq c|\epsilon_1|$ for some $c > 0$. Hence $\dot{\theta} \in \mathcal{S}(d_0^2 + \nu_0^2)$. Thus the adaptive law with the ϵ_1-modification guarantees that (ii) $\epsilon_1, \dot{\theta} \in \mathcal{S}(d_0^2 + \nu_0^2)$.

Now, if the modelling error disappears then $d_0 = 0$ and $\epsilon_1, \dot{\theta} \in \mathcal{S}(\nu_0^2)$. This shows that with the ϵ_1 modification, the ideal case L_2 properties are not recovered. Thus the ϵ_1 modification falls short of meeting the objective that originally served to motivate it.

Remark 6.3.1. The leakage modification discussed in this subsection can be derived by modifying the cost function $J(\theta) = \frac{\epsilon_1^2}{2} = \frac{(y-\theta u)^2}{2}$, used in the ideal case, to $J(\theta) = \frac{(y-\theta u)^2}{2} + w\frac{\theta^2}{2}$. The rationale for such a modification is that the new cost function penalizes the parameter estimate so that minimizing such a cost function would inhibit the phenomenon of parameter drift. Minimizing the new cost function using the gradient method, we obtain

$$\dot{\theta} = \gamma\epsilon_1 u - \gamma w\theta \tag{6.27}$$

which is identical to (6.11).

6.3.2 Parameter Projection

Instead of introducing a leakage modification into the adaptive law (6.3), another way of eliminating parameter drift would be to use the technique of parameter projection. Indeed, if the unknown parameter is apriori known to belong to a *bounded* convex set, then one can start the parameter estimation process inside this set and project the estimated parameters onto this set using the gradient projection method. This will ensure that the parameter estimate remains bounded. For the specific example (6.10) being considered here, suppose that $|\theta^*| \leq M_0$ where $M_0 > 0$ is known and consider the function $g(\theta) = \theta^2 - M_0^2$. Then the set $\mathcal{C} = \{\theta | g(\theta) \leq 0\}$ is a convex bounded set in R which also contains θ^*. Hence, applying the gradient projection method as in Example 4.5.1, we obtain

$$\dot{\theta} = \begin{cases} \gamma\epsilon_1 u & \text{if } |\theta| < M_0 \text{ or if } |\theta| = M_0 \text{ and } \theta\epsilon_1 u \leq 0 \\ 0 & \text{if } |\theta| = M_0 \text{ and } \theta\epsilon_1 u > 0 \end{cases} \tag{6.28}$$

$$\epsilon_1 = y - \theta u = -\tilde{\theta} u + \eta \tag{6.29}$$

By proceeding as in the proof of Theorem 4.5.1, we can show that the above adaptive law guarantees $|\theta(t)| \leq M_0$ provided $|\theta(0)| \leq M_0$. Now, let us analyze its other properties. Taking the time derivative of the Lyapunov-like function $V(\tilde{\theta}) = \frac{\tilde{\theta}^2}{2\gamma}$ along the solution of (6.28) we obtain

$$\dot{V} = \begin{cases} -\epsilon_1^2 + \epsilon_1\eta & \text{if } |\theta| < M_0 \text{ or if } |\theta| = M_0 \text{ and } \theta\epsilon_1 u \leq 0 \\ 0 & \text{if } |\theta| = M_0 \text{ and } \theta\epsilon_1 u > 0 \end{cases}$$

Let us now focus attention on the case when $\dot{V} = 0$, $|\theta| = M_0$ and $\theta\epsilon_1 u > 0$. Then, using the fact that $\epsilon_1 = -\tilde{\theta}u + \eta$, we can write $\dot{V} = 0 = -\epsilon_1^2 + \epsilon_1\eta - \tilde{\theta}\epsilon_1 u$.

$$\begin{aligned} \text{Again } \tilde{\theta}\epsilon_1 u &= \epsilon_1 u(\theta - \theta^*) \\ &= |\epsilon_1 u||\theta| - \epsilon_1 u\theta^* \text{ (since } \theta\epsilon_1 u > 0) \\ &= M_0|\epsilon_1 u| - \theta^*\epsilon_1 u \text{ (since } |\theta| = M_0) \\ &\geq (M_0 - |\theta^*|)|\epsilon_1 u| \end{aligned}$$

Since $M_0 > |\theta^*|$, it follows that $\tilde{\theta}\epsilon_1 u \geq 0$ which implies that $\dot{V} = 0 \leq -\epsilon_1^2 + \epsilon_1\eta$. Hence $\dot{V} \leq -\epsilon_1^2 + \epsilon_1\eta \;\forall\, t \geq 0$ so that by completing squares, we can write

$$\dot{V} \leq -\frac{\epsilon_1^2}{2} + \frac{d_0^2}{2} \; \forall \, t \geq 0.$$

Integrating the above, we obtain

$$\int_t^{t+T} \epsilon_1^2(\tau) d\tau \leq d_0^2 T + 2 \{V(t) - V(t+T)\}$$

$\forall \, t \geq 0$ and $\forall \, T \geq 0$. Thus $\epsilon_1 \in \mathcal{S}(d_0^2)$. Since, from (6.28), $|\dot{\theta}| \leq \gamma |\epsilon_1 u|$ and $u \in L_\infty$, it follows that $\dot{\theta} \in \mathcal{S}(d_0^2)$. Thus the projection modification guarantees properties which are very similar to those guaranteed by the switching-σ modification.

6.3.3 Dead Zone

The Dead Zone is yet another modification that can be used to robustify an adaptive law against the presence of modelling errors. The principal idea behind the dead-zone is to monitor the size of the estimation error and to adapt only when the signal-to-noise ratio (SNR) is high in the estimation error. Let us consider the adaptive law (6.3) rewritten in terms of the parameter error, i.e.

$$\dot{\tilde{\theta}} = \gamma \epsilon_1 u, \; \epsilon_1 = -\tilde{\theta} u + \eta \tag{6.30}$$

Since $\sup_t |\eta(t)| \leq d_0$, it follows that if $|\epsilon_1| \gg d_0$, then the signal $\tilde{\theta} u$ is dominant in ϵ_1 which means that the signal-to-noise ratio is high and one should continue with the adaptation. If, on the other hand, $|\epsilon_1| < d_0$, then η may be dominant in ϵ_1 so that adaptation should be stopped. The adaptive law (6.30) with the deadzone modification is given by

$$\dot{\theta} = \gamma u(\epsilon_1 + g), \; g = \begin{cases} 0 & \text{if } |\epsilon_1| \geq g_0 \\ -\epsilon_1 & \text{if } |\epsilon_1| < g_0 \end{cases} \tag{6.31}$$

$$\epsilon_1 = y - \theta u \tag{6.32}$$

where $g_0 > 0$ is a known strict upper bound for $|\eta(t)|$ i.e. $g_0 > d_0 \geq \sup_t |\eta(t)|$. The plot of the so called *deadzone function* $f(\epsilon_1) = \epsilon_1 + g$ versus ϵ_1 is shown in Fig. 6.2.

From Fig. 6.2, it is clear that the deadzone given by (6.31) is a discontinuous one which may lead to problems with the existence and uniqueness of solutions of the resulting differential equations [35]. To avoid any such problems, we can make the deadzone continuous as shown in Fig. 6.3 With the continuous deadzone, the adaptive law becomes

$$\dot{\theta} = \gamma u(\epsilon_1 + g), \; g = \begin{cases} g_0 & \text{if } \epsilon_1 < -g_0 \\ -g_0 & \text{if } \epsilon_1 > g_0 \\ -\epsilon_1 & \text{if } |\epsilon_1| \leq g_0 \end{cases} \tag{6.33}$$

which together with

$$\epsilon_1 = -\tilde{\theta} u + \eta \tag{6.34}$$

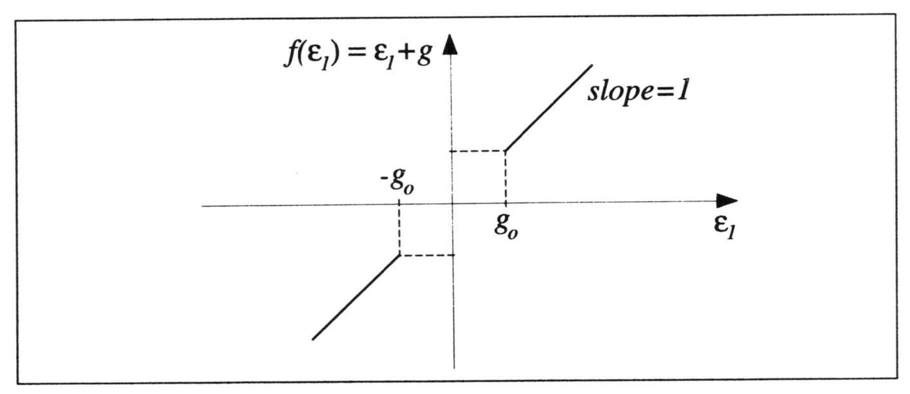

Fig. 6.2. Discontinuous Dead Zone Function

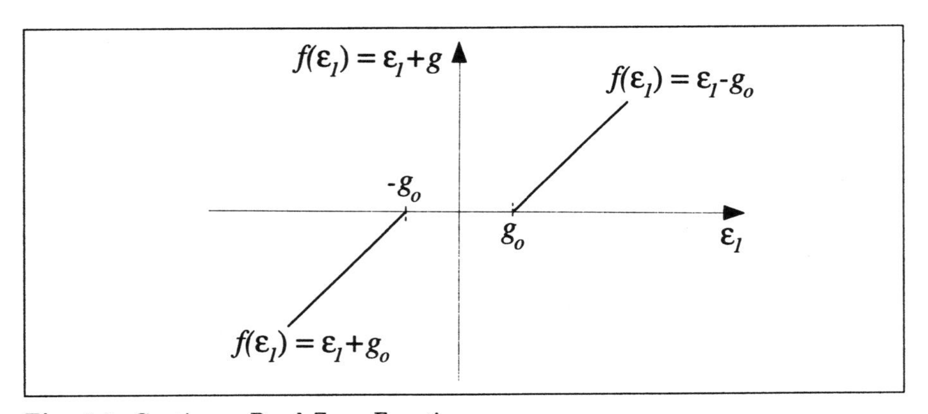

Fig. 6.3. Continuus Dead Zone Function

describe the stability properties of the modified adaptive law. Now consider the Lyapunov-like function $V(\tilde{\theta}) = \frac{\tilde{\theta}^2}{2\gamma}$ whose derivative along the solution of (6.33) is given by

$$\begin{aligned} \dot{V} &= \tilde{\theta}u(\epsilon_1 + g) \\ &= -(\epsilon_1 - \eta)(\epsilon_1 + g) \text{ (using (6.34)} \end{aligned}$$

From (6.33), we see that

$$(\epsilon_1 - \eta)(\epsilon_1 + g) = \begin{cases} (\epsilon_1 - \eta)(\epsilon_1 + g_0) > 0 & \text{if } \epsilon_1 < -g_0 < -|\eta| \\ (\epsilon_1 - \eta)(\epsilon_1 - g_0) > 0 & \text{if } \epsilon_1 > g_0 > |\eta| \\ 0 & \text{if } |\epsilon_1| \leq g_0 \end{cases}$$

Thus $(\epsilon_1 - \eta)(\epsilon_1 + g) \geq 0 \; \forall \, t \geq 0$ so that $\dot{V} \leq 0$ which implies that $V, \tilde{\theta}, \theta \in L_\infty$ and $\sqrt{(\epsilon_1 - \eta)(\epsilon_1 + g)} \in L_2$. From (6.34), (6.33), the boundedness of θ, u, η implies that $\epsilon_1, \dot{\theta} \in L_\infty$. Thus the property (i) can still be guaranteed with the dead-zone modification. We now proceed to examine how property (ii), i.e. the L_2 properties of $\dot{\theta}$ and ϵ_1 are affected. From (6.33) once again, we have

$$(\epsilon_1 + g)^2 = \begin{cases} (\epsilon_1 + g_0)^2 & \text{if } \epsilon_1 < -g_0 \\ (\epsilon_1 - g_0)^2 & \text{if } \epsilon_1 > g_0 \\ 0 & \text{if } |\epsilon_1| \leq g_0 \end{cases}$$

Furthermore,

$$(\epsilon_1 - \eta)(\epsilon_1 + g) = \begin{cases} (\epsilon_1 + g_0)^2 - (g_0 + \eta)(\epsilon_1 + g_0) & \text{if } \epsilon_1 < -g_0 \\ (\epsilon_1 - g_0)^2 + (g_0 - \eta)(\epsilon_1 - g_0) & \text{if } \epsilon_1 > g_0 \\ 0 & \text{if } |\epsilon_1| \leq g_0 \end{cases}$$

Thus $(\epsilon_1 - \eta)(\epsilon_1 + g) \geq (\epsilon_1 + g)^2$. Since $\sqrt{(\epsilon_1 - \eta)(\epsilon_1 + g)} \in L_2$, it follows that $(\epsilon_1 + g) \in L_2$ which, using (6.33) and the boundedness of u, implies that $\dot{\theta} \in L_2$. Hence, the dead-zone preserves the L_2 property of $\dot{\theta}$ despite the presence of the bounded disturbance η.

$$\begin{aligned} \text{Now } \dot{V} &= -(\epsilon_1 - \eta)(\epsilon_1 + g) \\ &= -\epsilon_1^2 - \epsilon_1 g + \epsilon_1 \eta + g\eta \\ &\leq -\epsilon_1^2 + |\epsilon_1| g_0 + |\epsilon_1| d_0 + d_0 g_0 \\ &\leq -\frac{\epsilon_1^2}{2} + \frac{(d_0 + g_0)^2}{2} + d_0 g_0 \text{ (by completing squares)} \\ &\leq -\frac{\epsilon_1^2}{2} + \frac{3}{2} d_0^2 + \frac{3}{2} g_0^2 \\ & \qquad \text{(using } (x+y)^2 \leq 2x^2 + 2y^2, xy \leq \tfrac{1}{2}(x^2 + y^2)) \end{aligned}$$

so that $\epsilon_1 \in \mathcal{S}(d_0^2 + g_0^2)$. We note that for the dead-zone modification, we cannot establish $\epsilon_1 \in L_2$ even when $\eta = 0$. In other words, the dead-zone modification achieves robustness at the expense of destroying some of the properties of the original adaptive law in the ideal case.

6.3.4 Dynamic Normalization

Let us now consider the problem of estimating θ^* in the model (6.10) where η is not necessarily bounded. If η satisfies $|\eta| \leq c_1|u| + c_2$ where c_1, c_2 are some non-negative constants, then we can divide both sides of (6.10) by m where $m = \sqrt{1+u^2}$ to obtain

$$\bar{y} = \theta^* \bar{u} + \bar{\eta} \tag{6.35}$$

where $\bar{x} \triangleq \frac{x}{m}$. Since $\bar{y}, \bar{u}, \bar{\eta} \in L_\infty$, the same procedure as before can be used to develop adaptive laws that are robust with respect to the bounded modelling error term $\bar{\eta}$. We would like to go a step further: consider the case where η is not necessarily bounded from above by $|u|$ but is related to u through some transfer function, e.g.

$$y = \theta^*(1 + \Delta_m(s))[u] + d \tag{6.36}$$

where $\Delta_m(s)$ is an *unknown* multiplicative perturbation that is strictly proper and stable, and d is an unknown bounded disturbance. Equation (6.36) above can be rewritten as

$$\begin{aligned} y &= \theta^* u + \eta & (6.37)\\ \text{where } \eta &= \theta^* \Delta_m(s)[u] + d & (6.38) \end{aligned}$$

To apply the procedure of the previous subsections, we need to find a normalizing signal m that allows us to rewrite the above equation as

$$\bar{y} = \theta^* \bar{u} + \bar{\eta} \tag{6.39}$$

where $\bar{y} \triangleq \frac{y}{m}$, $\bar{u} \triangleq \frac{u}{m}$, $\bar{\eta} \triangleq \frac{\eta}{m}$ and $\bar{y}, \bar{u}, \bar{\eta} \in L_\infty$. In this case, the choice $m^2 = 1 + u^2$ will not guarantee the boundedness of $\frac{\eta}{m}$. Hence, we proceed as follows. Now, from (6.38), we see that the relationship

$$|\eta(t)| \leq |\theta^*| ||\Delta_m(s)||_2^\delta ||u_t||_2^\delta + |d(t)|$$

holds for any $\delta \geq 0$ provided $\Delta_m(s)$ is strictly proper and analytic in Re$[s] \geq -\frac{\delta}{2}$ (See Chapter 2, Properties of the L_2^δ norm, Lemma 2.3.3(ii)). If we now assume that in addition to being strictly proper, $\Delta_m(s)$ satisfies the following assumption:
A1. $\Delta_m(s)$ is analytic in Re$[s] \geq -\frac{\delta_0}{2}$ for some known $\delta_0 > 0$, then we have

$$|\eta(t)| \leq \mu_0 ||u_t||_2^{\delta_0} + d_0 \tag{6.40}$$

where $\mu_0 = |\theta^*|.||\Delta_m(s)||_2^{\delta_0}$ and d_0 is an upper bound for $|d(t)|$. Because μ_0, d_0 are constants, (6.40) motivates the normalizing signal m given by $m^2 = 1 + u^2 + \left(||u_t||_2^{\delta_0}\right)^2$. This signal m may be generated by the equations

$$\begin{aligned} \dot{m}_s &= -\delta_0 m_s + u^2,\ m_s(0) = 0 &(6.41)\\ n_s^2 &= m_s,\ m^2 = 1 + u^2 + n_s^2 &(6.42)\end{aligned}$$

We refer to $m_s = n_s^2$ as the *dynamic normalizing signal* in order to distinguish it from the *static* one given by $m^2 = 1 + u^2$ which was used for the ideal case example in Chapter 4, Section 4.4.1. Now we can start with (6.39) and proceed as in the previous subsections to design and analyze adaptive laws for the model (6.10) when the modelling error term η is not necessarily bounded. For instance, using the leakage modification, we obtain the adaptive law

$$\dot{\theta} = \gamma \bar{\epsilon}_1 \bar{u} - \gamma w \theta \tag{6.43}$$

where $\bar{\epsilon}_1 = \bar{y} - \hat{\bar{y}} = \frac{y - \theta u}{m}$, $\hat{\bar{y}} = \theta \bar{u}$ and w is the leakage term to be chosen. As in Section 4.4.1, the adaptive law (6.43) can be written as

$$\dot{\theta} = \gamma \epsilon u - \gamma w \theta \tag{6.44}$$

where $\epsilon = \frac{y - \theta u}{m^2}$ is the *normalized estimation error.*

The analysis of the normalized adaptive law (6.44) is very similar to that of the unnormalized adaptive law (6.11). As before, we consider the Lyapunov-like function $V(\tilde{\theta}) = \frac{\tilde{\theta}^2}{2\gamma}$, $\tilde{\theta} \triangleq \theta - \theta^*$. Computing its derivative along the solution of (6.44), we obtain

$$\begin{aligned} \dot{V} &= \tilde{\theta}[\epsilon u - w\theta]\\ &= \epsilon \tilde{\theta} u - w \tilde{\theta} \theta\\ &= \epsilon(-\epsilon m^2 + \eta) - w\tilde{\theta}\theta \ (\text{since } \epsilon = \tfrac{y - \theta u}{m^2} = \tfrac{-\tilde{\theta} u + \eta}{m^2}\ )\\ &= -\epsilon^2 m^2 + \epsilon \eta - w \tilde{\theta}\theta \end{aligned}$$

As in Section 6.3.1, we now consider a specific choice of w say $w = \sigma$. Then we have

$$\begin{aligned} \dot{V} &= -\epsilon^2 m^2 + \epsilon\eta - \sigma\tilde{\theta}\theta\\ &\leq -\epsilon^2 m^2 + |\epsilon m| \frac{|\eta|}{m} - \frac{\sigma}{2}|\tilde{\theta}|^2 + \frac{\sigma}{2}|\theta^*|^2 \ (\text{using (6.16)})\\ &\leq -\frac{\epsilon^2 m^2}{2} + \frac{\eta^2}{2m^2} - \frac{\sigma}{2}\tilde{\theta}^2 + \frac{\sigma}{2}\theta^{*2} \qquad (6.45) \end{aligned}$$

(completing squares using the first two terms)

Since $\frac{\eta}{m} \in L_\infty$, it follows that $V, \tilde{\theta} \in L_\infty$. Furthermore, integrating both sides of (6.45) it follows that $\epsilon m \in \mathcal{S}\left(\frac{\eta^2}{m^2} + \sigma\right)$ which in turn implies that $\epsilon \in \mathcal{S}\left(\frac{\eta^2}{m^2} + \sigma\right)$. Since $\dot{\tilde{\theta}} = \gamma \epsilon m \frac{u}{m} - \gamma\sigma\theta$ and $\frac{u}{m} \in L_\infty$, we can proceed as in Section 6.3.1 to show that $\dot{\tilde{\theta}} \in \mathcal{S}\left(\frac{\eta^2}{m^2} + \sigma\right)$. The boundedness of $\tilde{\theta}$, $\frac{u}{m}$, $\frac{\eta}{m}$ also implies that $\epsilon, \epsilon m, \dot{\theta} \in L_\infty$. Thus, with $w = \sigma$, we have (i) $\epsilon, \epsilon m, \theta, \dot{\theta} \in L_\infty$ and (ii) $\epsilon, \epsilon m, \dot{\theta} \in \mathcal{S}\left(\frac{\eta^2}{m^2} + \sigma\right)$. Similar properties can be established when dynamic normalization is used in conjunction with any of the other robustness modifications discussed in Sections 6.3.1, 6.3.2 or 6.3.3.

6.4 Robust Adaptive Laws

In Chapter 4 we considered the design and analysis of adaptive laws for estimating the unknown parameters of a general nth order plant. Such a design and analysis was carried out under the assumption that there are no modelling errors, i.e. the plant is a perfectly modelled one. Since no physical system can be modelled perfectly, such an assumption is unrealistic, to say the very least. Furthermore, the examples of Section 6.2 have shown that any violation of this perfect modelling assumption can potentially lead to catastrophic results. Our objective in this section, therefore, is to discuss the various modifications that can be used to robustify the adaptive laws in the presence of modelling errors. These modifications were already introduced in the last section, where they were illustrated on a simple scalar example. Here we will extend those results to handle the general case. To this end, we first consider the representation of a general plant with modelling errors in a form that is suitable for robust parameter estimation.

6.4.1 Parametric Model with Modelling Error

Let us start with the plant[1]

$$y = \frac{Z_0(s)}{R_0(s)}(1 + \mu\Delta_m(s))[u] + d \tag{6.46}$$

where u, y are the input and output signals of the plant; $R_0(s)$ is a monic polynomial of degree n; $Z_0(s)$ is a polynomial of degree l with $l < n$; $\frac{Z_0(s)}{R_0(s)}$ represents the transfer function of the modelled part of the plant; $\mu\Delta_m(s)$, $\mu > 0$ is a stable multiplicative uncertainty and $d(t)$ is a bounded disturbance. We now proceed as in Section 4.3 to express the plant (6.46) in a form suitable for robust parameter estimation. Suppose that $R_0(s) = s^n + a_1 s^{n-1} + a_2 s^{n-2} + \cdots + a_{n-1}s + a_n$, $Z_0(s) = b_0 s^l + b_1 s^{l-1} + \cdots + b_l$ and let $\Lambda(s) = s^n + \lambda_1 s^{n-1} + \lambda_2 s^{n-2} + \cdots + \lambda_n$ be an arbitrary monic Hurwitz polynomial of degree n. Then rewriting (6.46) as

$$R_0(s)[y] = Z_0(s)[u] + \mu\Delta_m(s)Z_0(s)[u] + R_0(s)[d] \tag{6.47}$$

and filtering both sides of (6.47) with the stable filter $\frac{1}{\Lambda(s)}$, we obtain

$$\begin{aligned}
&\frac{s^n}{\Lambda(s)}[y] + a_1\frac{s^{n-1}}{\Lambda(s)}[y] + a_2\frac{s^{n-2}}{\Lambda(s)}[y] + \cdots + a_n\frac{1}{\Lambda(s)}[y] = \\
&b_0\frac{s^l}{\Lambda(s)}[u] + b_1\frac{s^{l-1}}{\Lambda(s)}[u] + \cdots + b_l\frac{1}{\Lambda(s)}[u] \\
&+\mu\frac{\Delta_m(s)Z_0(s)}{\Lambda(s)}[u] + \frac{R_0(s)}{\Lambda(s)}[d]
\end{aligned} \tag{6.48}$$

[1] Without any loss of generality we assume that the unmodelled dynamics are of the multiplicative type; other uncertainty types can be handled as in [19].

Defining

$$\begin{aligned}\phi &= \left[\frac{s^{n-1}}{\Lambda(s)}[y], \frac{s^{n-2}}{\Lambda(s)}[y], \cdots, \frac{1}{\Lambda(s)}[y], \frac{s^{l}}{\Lambda(s)}[u], \frac{s^{l-1}}{\Lambda(s)}[u], \cdots, \right. \\ &\quad \left. \frac{1}{\Lambda(s)}[u]\right]^T \quad (6.49)\\ \eta &= \mu\frac{\Delta_m(s)Z_0(s)}{\Lambda(s)}[u] + \frac{R_0(s)}{\Lambda(s)}[d] \quad (6.50)\end{aligned}$$

we can express (6.48) as

$$z = \theta^{*T}\phi + \eta \quad (6.51)$$

where either

$$\begin{aligned} z &= \frac{s^n}{\Lambda(s)}[y],\ \theta^* = [-a_1, -a_2, \cdots, -a_n, b_0, b_1, \cdots, b_l]^T \quad (6.52)\\ \text{or } z &= y,\ \theta^* = [\lambda_1 - a_1, \lambda_2 - a_2, \cdots, \lambda_n - a_n, b_0, b_1, \cdots, b_l]^T \quad (6.53)\end{aligned}$$

Equation (6.51) above is referred to as the *linear parametric model with modelling error* and is suitable for robust parameter estimation. In this model, the signals z and ϕ are available for measurement, θ^* is the unknown parameter vector to be estimated and η is an unknown modelling error term. We will show that the adaptive laws of Chapter 4 developed for the parametric model (6.51) with $\eta = 0$ can be modified to handle the case where $\eta \neq 0$ is of the form (6.50).

6.4.2 Gradient Algorithm with Leakage

Consider the parametric model (6.51). The procedure for designing an adaptive law for estimating on-line the constant vector θ^* in the presence of the unknown modelling error term η is very similar to that in the ideal case ($\eta = 0$) presented in Chapter 4. As in the ideal case, the predicted value $\hat{z}$ of z based on the estimate θ of θ^* and the normalized estimation error ϵ are generated as

$$\begin{aligned}\hat{z} &= \theta^T\phi \quad (6.54)\\ \epsilon &= \frac{z - \hat{z}}{m^2} \quad (6.55)\end{aligned}$$

where $m^2 = 1 + n_s^2$ is the normalizing signal to be designed. From (6.51), (6.54), (6.55), we obtain

$$\epsilon = \frac{-\tilde{\theta}^T\phi + \eta}{m^2} \quad (6.56)$$

where $\tilde{\theta} \triangleq \theta - \theta^*$. If we now design m so that
(A1) $\frac{\phi}{m}, \frac{\eta}{m} \in L_\infty$

then the signal ϵm is a reasonable measure for the parameter error $\tilde{\theta}$ since for any piecewise continuous signals ϕ and η, large ϵm implies large $\tilde{\theta}$.

Design of the Normalizing Signal

Assume that $\frac{\Delta_m(s)Z_0(s)}{\Lambda(s)}$ is strictly proper and both $\frac{1}{\Lambda(s)}$, $\frac{\Delta_m(s)Z_0(s)}{\Lambda(s)}$ are analytic in $\mathrm{Re}[s] \geq -\frac{\delta_0}{2}$. Then according to Lemma 2.3.3, we have

$$\begin{aligned} |\eta(t)| &\leq \left\| \frac{\mu\Delta_m(s)Z_0(s)}{\Lambda(s)} \right\|_2^{\delta_0} \|u_t\|_2^{\delta_0} + \left| \left(\frac{R_0(s)}{\Lambda(s)}[d] \right)(t) \right| \\ \text{and } |\phi(t)| &\leq c(\|u_t\|_2^{\delta_0} + \|y_t\|_2^{\delta_0}) \end{aligned}$$

for some $c > 0$. This motivates the following normalizing signal

$$m^2 = 1 + n_s^2, \; n_s^2 = m_s \tag{6.57}$$

$$\dot{m}_s = -\delta_0 m_s + u^2 + y^2, \; m_s(0) = 0 \tag{6.58}$$

that satisfies (A1). The gradient algorithm with leakage can now be derived by minimizing the cost function $J = \frac{(z-\theta^T\phi)^2}{2m^2} + w(t)\theta^T\theta$ with respect to θ where $w(t) \geq 0$ is a design function that acts as a weighting coefficient. Note that the cost function here is the one considered in Section 4.4.3 augmented with the $w(t)\theta^T\theta$ term. Using the gradient method, we obtain

$$\dot{\theta} = \gamma\epsilon\phi - \gamma w(t)\theta \tag{6.59}$$

$$\text{where } \epsilon = \frac{z - \theta^T\phi}{m^2} \tag{6.60}$$

Thus the weighting $w(t)$ in the cost function appears as a leakage term in the adaptive law (6.59). The following theorems describe the stability properties of the adaptive law (6.59) for different choices of $w(t)$.

Theorem 6.4.1. (Fixed σ modification) *Let*

$$w(t) = \sigma > 0, \; \forall\, t \geq 0$$

where σ is a small constant. Then the adaptive law (6.59) guarantees that
(i) $\epsilon, \epsilon n_s, \theta, \dot{\theta} \in L_\infty$
(ii) $\epsilon, \epsilon n_s, \dot{\theta} \in \mathcal{S}\left(\frac{\eta^2}{m^2} + \sigma\right)$.

Proof. With the choice $w(t) = \sigma$, the adaptive law (6.59) can be rewritten as

$$\dot{\tilde{\theta}} = \gamma\epsilon\phi - \gamma\sigma\theta, \; \tilde{\theta} \triangleq \theta - \theta^* \tag{6.61}$$

and using (6.51), (6.60), the normalized estimation error ϵ satisfies

$$\epsilon = \frac{-\tilde{\theta}^T\phi + \eta}{m^2} \tag{6.62}$$

Now consider the Lyapunov-like function $V = \frac{\tilde{\theta}^T\tilde{\theta}}{2\gamma}$ whose derivative along the solution of (6.61) is given by

$$\begin{aligned}
\dot{V} &= \epsilon\tilde{\theta}^T\phi - \sigma\tilde{\theta}^T\theta \\
&= \epsilon(-\epsilon m^2 + \eta) - \sigma\tilde{\theta}^T\theta \text{ (using (6.62))} \\
&= -\epsilon^2 m^2 + \epsilon\eta - \sigma\tilde{\theta}^T\theta \\
&\leq -\epsilon^2 m^2 + |\epsilon m|\frac{|\eta|}{m} - \sigma\tilde{\theta}^T\theta
\end{aligned} \tag{6.63}$$

Now by a simple completion of squares, we obtain

$$\begin{aligned}
-\epsilon^2 m^2 + |\epsilon m|\frac{|\eta|}{m} &= -\frac{\epsilon^2 m^2}{2} - \frac{1}{2}[(\epsilon m)^2 - 2|\epsilon m|\frac{|\eta|}{m}] \\
&= -\frac{\epsilon^2 m^2}{2} - \frac{1}{2}[|\epsilon m| - \frac{|\eta|}{m}]^2 + \frac{\eta^2}{2m^2} \\
&\leq -\frac{\epsilon^2 m^2}{2} + \frac{\eta^2}{2m^2}
\end{aligned} \tag{6.64}$$

$$\begin{aligned}
\text{Also } -\sigma\tilde{\theta}^T\theta &= -\sigma\tilde{\theta}^T(\tilde{\theta} + \theta^*) \\
&\leq -\sigma|\tilde{\theta}|^2 + \sigma|\tilde{\theta}||\theta^*| \\
&= -\frac{\sigma}{2}|\tilde{\theta}|^2 - \frac{\sigma}{2}(|\tilde{\theta}|^2 - 2|\tilde{\theta}||\theta^*|) \\
&= -\frac{\sigma}{2}|\tilde{\theta}|^2 - \frac{\sigma}{2}(|\tilde{\theta}| - |\theta^*|)^2 + \frac{\sigma}{2}|\theta^*|^2 \\
&\leq -\frac{\sigma}{2}|\tilde{\theta}|^2 + \frac{\sigma}{2}|\theta^*|^2
\end{aligned} \tag{6.65}$$

Thus $\dot{V} \leq -\frac{\epsilon^2 m^2}{2} - \frac{\sigma}{2}|\tilde{\theta}|^2 + \frac{\eta^2}{2m^2} + \frac{\sigma}{2}|\theta^*|^2$. Adding and subtracting the term αV where $\alpha > 0$ is arbitrary, we obtain

$$\dot{V} \leq -\alpha V - \frac{\epsilon^2 m^2}{2} - (\sigma - \frac{\alpha}{\gamma})\frac{|\tilde{\theta}|^2}{2} + \frac{\eta^2}{2m^2} + \frac{\sigma}{2}|\theta^*|^2.$$

Choosing $0 < \alpha \leq \sigma\gamma$, we have

$$\dot{V} \leq -\alpha V + \frac{\eta^2}{2m^2} + \frac{\sigma}{2}|\theta^*|^2 \tag{6.66}$$

Since $\frac{\eta}{m} \in L_\infty$, it follows that for $V \geq V_0 \triangleq \frac{1}{2\alpha}(\sup_t \frac{\eta^2}{m^2} + \sigma|\theta^*|^2)$, $\dot{V} \leq 0$ so that $\tilde{\theta}, \theta \in L_\infty$. From (6.62), using the fact that $\frac{\phi}{m}, \frac{\eta}{m} \in L_\infty$, it now follows that $\epsilon, \epsilon n_s \in L_\infty$. Furthermore, from (6.61), we obtain

$$|\dot{\theta}| \leq \gamma|\epsilon m|\frac{|\phi|}{m} + \gamma\sigma|\theta|$$

Since $\frac{\phi}{m}, \epsilon m, \theta \in L_\infty$, we conclude that $\dot{\theta} \in L_\infty$ and this completes the proof of property (i).

Let us now turn to the proof of property (ii). We have

$$\begin{aligned} -\sigma\tilde{\theta}^T\theta &= -\sigma(\theta-\theta^*)^T\theta \\ &\le -\frac{\sigma}{2}|\theta|^2+\frac{\sigma}{2}|\theta^*|^2 \\ &\quad \text{(completing squares as in the derivation of (6.65))} \end{aligned}$$

so that from (6.63) and (6.64) we obtain

$$\dot{V} \le -\frac{\epsilon^2 m^2}{2}-\frac{\sigma}{2}|\theta|^2+\frac{\eta^2}{2m^2}+\frac{\sigma}{2}|\theta^*|^2.$$

Integrating both sides of the above inequality, we obtain

$$\begin{aligned} \int_t^{t+T}\epsilon^2 m^2 d\tau+\int_t^{t+T}\sigma|\theta|^2 d\tau &\le \int_t^{t+T}\left(\frac{\eta^2}{m^2}+\sigma|\theta^*|^2\right)d\tau \\ &\quad +2[V(t)-V(t+T)] \end{aligned}$$

which shows that $\epsilon m, \sqrt{\sigma}|\theta| \in \mathcal{S}\left(\frac{\eta^2}{m^2}+\sigma\right)$. Since $\epsilon^2 m^2=\epsilon^2+\epsilon^2 n_s^2$, it follows that $\epsilon, \epsilon n_s \in \mathcal{S}\left(\frac{\eta^2}{m^2}+\sigma\right)$. Now from (6.61), using the identity $(a+b)^2 \le 2a^2+2b^2$, we obtain

$$|\dot{\theta}|^2 \le 2\gamma^2\epsilon^2 m^2\frac{|\phi|^2}{m^2}+2\gamma^2\sigma^2|\theta|^2.$$

Since $\frac{|\phi|}{m}, \sigma \in L_\infty$, if follows that $\dot{\theta} \in \mathcal{S}\left(\frac{\eta^2}{m^2}+\sigma\right)$. This completes the proof.

Theorem 6.4.2. (Switching σ). *Let*

$$w(t)=\sigma_s,\ \sigma_s=\begin{cases} 0 & \text{if } |\theta(t)| \le M_0 \\ \sigma_0\left(\frac{|\theta(t)|}{M_0}-1\right) & \text{if } M_0 < |\theta(t)| \le 2M_0 \\ \sigma_0 & \text{if } |\theta(t)| > 2M_0 \end{cases} \tag{6.67}$$

and σ_0, M_0 are design constants with $M_0 > |\theta^*|$ and $\sigma_0 > 0$. Then the adaptive law (6.59) guarantees that
(i) $\epsilon, \epsilon n_s, \theta, \dot{\theta} \in L_\infty$
(ii) $\epsilon, \epsilon n_s, \dot{\theta} \in \mathcal{S}\left(\frac{\eta^2}{m^2}\right)$.
(iii) In the absence of modelling errors, i.e. when $\eta=0$, property (ii) can be replaced with $(ii)'$ $\epsilon, \epsilon n_s, \dot{\theta} \in L_2$.

Proof. With the choice $w(t)=\sigma_s$, the adaptive law (6.59) can be rewritten as

$$\dot{\tilde{\theta}}=\gamma\epsilon\phi-\gamma\sigma_s\theta,\ \tilde{\theta} \triangleq \theta-\theta^* \tag{6.68}$$

Starting with $V=\frac{\tilde{\theta}^T\tilde{\theta}}{2\gamma}$ and proceeding as in the derivation of (6.63), (6.64), we obtain

$$\dot{V} \leq -\frac{\epsilon^2 m^2}{2} - \sigma_s \tilde{\theta}^T \theta + \frac{\eta^2}{2m^2} \tag{6.69}$$

$$\begin{aligned}
\text{Now } \sigma_s \tilde{\theta}^T \theta &= \sigma_s (\theta - \theta^*)^T \theta \\
&\geq \sigma_s |\theta|^2 - \sigma_s |\theta^*||\theta| \\
&= \sigma_s |\theta|(|\theta| - M_0 + M_0 - |\theta^*|) \\
&= \sigma_s |\theta|(|\theta| - M_0) + \sigma_s |\theta|(M_0 - |\theta^*|) \\
&\geq 0 \text{ (by (6.67) and the fact that } M_0 > |\theta^*|)
\end{aligned} \tag{6.70}$$

Thus the switching σ can only make $\dot{V}$ more negative. We next show that the switching σ modification can guarantee similar properties as the fixed σ modification. Now $-\sigma_s \tilde{\theta}^T \theta = -\sigma_0 \tilde{\theta}^T \theta + (\sigma_0 - \sigma_s) \tilde{\theta}^T \theta$. Furthermore, from (6.67)

$$\sigma_0 - \sigma_s = \begin{cases} \sigma_0 & \text{if } |\theta| \leq M_0 \\ \sigma_0 \left(2 - \frac{|\theta|}{M_0}\right) & \text{if } M_0 < |\theta| \leq 2M_0 \\ 0 & \text{if } |\theta| > 2M_0 \end{cases}$$

Since $|\tilde{\theta}| = |\theta - \theta^*| \leq |\theta| + |\theta^*| \leq |\theta| + M_0$, it follows that

$$(\sigma_0 - \sigma_s)\tilde{\theta}^T \theta \leq \begin{cases} 2\sigma_0 M_0^2 & \text{if } |\theta| \leq M_0 \\ 6\sigma_0 M_0^2 & \text{if } M_0 < |\theta| \leq 2M_0 \\ 0 & \text{if } |\theta| > 2M_0 \end{cases}$$

Thus $-\sigma_s \tilde{\theta}^T \theta \leq -\sigma_0 \tilde{\theta}^T \theta + 6\sigma_0 M_0^2$ so that from (6.69)

$$\dot{V} \leq -\frac{\epsilon^2 m^2}{2} - \sigma_0 \tilde{\theta}^T \theta + 6\sigma_0 M_0^2 + \frac{\eta^2}{2m^2}.$$

From the above inequality, it follows that the boundedness of $\tilde{\theta}, \theta, \epsilon, \epsilon n_s, \dot{\theta}$ can be concluded using the same arguments as in the case of the fixed σ modification. This completes the proof of property (i).

We next turn to the proof of property (ii). Now, integrating (6.69), we have

$$2\int_t^{t+T} \sigma_s \tilde{\theta}^T \theta d\tau + \int_t^{t+T} \epsilon^2 m^2 d\tau \leq \int_t^{t+T} \frac{\eta^2}{m^2} d\tau + c_1$$

where $c_1 = 2\sup_{t\geq 0,\, T\geq 0}[V(t) - V(t+T)]$. Since $\sigma_s \tilde{\theta}^T \theta \geq 0$, it follows that $\sqrt{\sigma_s \tilde{\theta}^T \theta}, \epsilon m \in \mathcal{S}\left(\frac{\eta^2}{m^2}\right)$. Since $\epsilon^2 m^2 = \epsilon^2 + \epsilon^2 n_s^2$ and $\epsilon m \in \mathcal{S}\left(\frac{\eta^2}{m^2}\right)$, we conclude that $\epsilon, \epsilon n_s \in \mathcal{S}\left(\frac{\eta^2}{m^2}\right)$. Again, from (6.70), we have

$$\sigma_s \tilde{\theta}^T \theta \geq \sigma_s |\theta|(M_0 - |\theta^*|)$$

which implies that

$$\sigma_s^2|\theta|^2 \leq \frac{(\sigma_s\tilde{\theta}^T\theta)^2}{(M_0 - |\theta^*|)^2} \leq c\sigma_s\tilde{\theta}^T\theta$$

where $c > 0$ is some constant. Hence it follows that $\sigma_s|\theta| \in \mathcal{S}\left(\frac{\eta^2}{m^2}\right)$. Furthermore, from (6.68), we have

$$|\dot{\theta}|^2 \leq 2\gamma^2\epsilon^2 m^2\frac{|\phi|^2}{m^2} + 2\gamma^2\sigma_s^2|\theta|^2.$$

Since $\frac{\phi}{m} \in L_\infty$, $\epsilon m, \sigma_s|\theta| \in \mathcal{S}\left(\frac{\eta^2}{m^2}\right)$, we conclude that $\dot{\theta} \in \mathcal{S}\left(\frac{\eta^2}{m^2}\right)$. This completes the proof of property (ii). The proof for property (iii) follows from (6.69) by setting $\eta = 0$, using $-\sigma_s\tilde{\theta}^T\theta \leq 0$ and repeating the above calculations with $\eta = 0$.

Theorem 6.4.3. (ϵ Modification). *Let*

$$w(t) = |\epsilon m|\nu_0$$

where $\nu_0 > 0$ is a design constant. Then the adaptive law (6.59) with $w(t) = |\epsilon m|\nu_0$ guarantees that
(i) $\epsilon, \epsilon n_s, \theta, \dot{\theta} \in L_\infty$
(ii) $\epsilon, \epsilon n_s, \dot{\theta} \in \mathcal{S}\left(\frac{\eta^2}{m^2} + \nu_0\right)$

Proof. With the choice $w(t) = |\epsilon m|\nu_0$, the adaptive law (6.59) can be rewritten as

$$\dot{\tilde{\theta}} = \gamma\epsilon\phi - \gamma|\epsilon m|\nu_0\theta, \; \tilde{\theta} \triangleq \theta - \theta^* \tag{6.71}$$

where the normalized estimation error ϵ satisfies

$$\epsilon = \frac{-\tilde{\theta}^T\phi + \eta}{m^2} \tag{6.72}$$

Now consider the Lyapunov-like function $V(\tilde{\theta}) = \frac{\tilde{\theta}^T\tilde{\theta}}{2\gamma}$ and evaluate its derivative along the solution of (6.71) to obtain

$$\begin{aligned} \dot{V} &= \epsilon\tilde{\theta}^T\phi - |\epsilon m|\nu_0\tilde{\theta}^T\theta \\ &= \epsilon[-\epsilon m^2 + \eta] - |\epsilon m|\nu_0\tilde{\theta}^T\theta \text{ (using (6.72))} \\ &\leq -|\epsilon m|(|\epsilon m| + \nu_0\frac{|\tilde{\theta}|^2}{2} - \nu_0\frac{|\theta^*|^2}{2} - \frac{|\eta|}{m}) \end{aligned} \tag{6.73}$$

where the above inequality is obtained by using

$$-\nu_0\tilde{\theta}^T\theta \leq -\nu_0\frac{|\tilde{\theta}|^2}{2} + \nu_0\frac{|\theta^*|^2}{2}.$$

Since $\frac{\eta}{m} \in L_\infty$, it is now clear that for $\nu_0\frac{|\tilde{\theta}|^2}{2} \geq \nu_0\frac{|\theta^*|^2}{2} + \sup_{t\geq 0}\frac{|\eta|}{m}$, i.e. for $V \geq V_0 \triangleq \frac{1}{\gamma\nu_0}(\nu_0\frac{|\theta^*|^2}{2} + \sup_{t\geq 0}\frac{|\eta|}{m})$, we have $\dot{V} \leq 0$ which implies that V and, therefore, $\theta, \tilde{\theta} \in L_\infty$. From (6.72), using the fact that $\frac{\phi}{m}, \frac{\eta}{m} \in L_\infty$, it now follows that $\epsilon, \epsilon n_s \in L_\infty$. Furthermore from (6.71), we obtain

$$|\dot{\theta}| \leq \gamma|\epsilon m|\frac{|\phi|}{m} + \gamma|\epsilon m|\nu_0|\theta|.$$

Since $\frac{|\phi|}{m}, \epsilon m, \theta \in L_\infty$, we conclude that $\dot{\theta} \in L_\infty$ and this completes the proof of property (i).

We now turn to the proof of property (ii). From (6.73)

$$\begin{aligned}
\dot{V} &= -\epsilon^2 m^2 + \epsilon m\frac{\eta}{m} - |\epsilon m|\nu_0\tilde{\theta}^T(\tilde{\theta}+\theta^*) \\
&\leq -\epsilon^2 m^2 + |\epsilon m|\frac{|\eta|}{m} - |\epsilon m|\nu_0|\tilde{\theta}|^2 - |\epsilon m|\nu_0\tilde{\theta}^T\theta^* \\
&\leq -\frac{\epsilon^2 m^2}{2} + \frac{\eta^2}{2m^2} - |\epsilon m|\frac{\nu_0}{2}|\tilde{\theta}|^2 + |\epsilon m|\frac{\nu_0}{2}|\theta^*|^2 \\
&\qquad \text{(completing squares and using } \pm\tilde{\theta}^T\theta^* \leq \tfrac{|\theta^*|^2+|\tilde{\theta}|^2}{2}\text{)} \\
&\leq -\frac{\epsilon^2 m^2}{4} - \frac{\epsilon^2 m^2}{4} + |\epsilon m|\frac{\nu_0}{2}|\theta^*|^2 + \frac{\eta^2}{2m^2} \\
&\leq -\frac{\epsilon^2 m^2}{4} + \nu_0^2\frac{|\theta^*|^4}{4} + \frac{\eta^2}{2m^2} \text{ (completing squares)}
\end{aligned}$$

Now integrating on both sides, we establish that $\epsilon m \in \mathcal{S}\left(\frac{\eta^2}{m^2} + \nu_0\right)$. Since $\epsilon^2 m^2 = \epsilon^2 + \epsilon^2 n_s^2$ and $\epsilon m \in \mathcal{S}\left(\frac{\eta^2}{m^2} + \nu_0\right)$, we conclude that $\epsilon, \epsilon n_s \in \mathcal{S}\left(\frac{\eta^2}{m^2} + \nu_0\right)$. Furthermore from (6.71), we have

$$|\dot{\theta}|^2 \leq 2\gamma^2\epsilon^2 m^2\frac{|\phi|^2}{m^2} + 2\gamma^2|\epsilon m|^2\nu_0^2|\theta|^2.$$

Since $\frac{\phi}{m}, \theta \in L_\infty$ and $\epsilon m \in \mathcal{S}\left(\frac{\eta^2}{m^2} + \nu_0\right)$, we conclude that $\dot{\theta} \in \mathcal{S}\left(\frac{\eta^2}{m^2} + \nu_0\right)$. This completes the proof of property (ii).

6.4.3 Parameter Projection

The two crucial techniques that we used in Section 6.4.2 to develop robust adaptive laws are the dynamic normalization and leakage. The normalization guarantees that the normalized modelling error $\frac{\eta}{m}$ is bounded and therefore enters the adaptive law as a bounded disturbance input. Since a bounded disturbance may cause parameter drift, the leakage modification is used to guarantee bounded parameter estimates. An alternative way to guarantee bounded parameter estimates is to use projection to constrain the parameter estimates to lie inside some known convex bounded set that contains the unknown θ^*. Adaptive laws with projection have already been introduced and analyzed in Chapter 4 for the ideal case. In this subsection, our objective is to examine how gradient adaptive laws with projection retain certain desirable properties even in the presence of modelling errors.

Consider the parametric model (6.51) and suppose that to avoid parameter drift in θ, the estimate of θ^*, we constrain θ to lie inside a convex *bounded* set $\mathcal{C}$ that contains θ^*. Furthermore, suppose that the set $\mathcal{C}$ is given by $\mathcal{C} = \{\theta \in R^{n+l+1} \mid g(\theta) \leq 0\}$ where $g : R^{n+l+1} \mapsto R$. As an example, we could have $g(\theta) = \theta^T\theta - M_0^2$ where M_0 is chosen such that $M_0 \geq |\theta^*|$.

Following the results of Chapter 4, the normalized gradient adaptive law with projection becomes

$$\dot{\theta} = \begin{cases} \gamma\epsilon\phi & \text{if } \theta \in \mathcal{C}^0 \text{ or if } \theta \in \delta\mathcal{C} \text{ and } (\gamma\epsilon\phi)^T\nabla g \leq 0 \\ \gamma\epsilon\phi - \frac{(\nabla g \nabla g^T)}{\nabla g^T \nabla g}\gamma\epsilon\phi & \text{otherwise} \end{cases} \tag{6.74}$$

where $\theta(0)$ is chosen so that $\theta(0) \in \mathcal{C}$ and $\epsilon = \frac{z-\theta^T\phi}{m^2}$, $\gamma > 0$.

The stability properties of (6.74) for estimating θ^* in (6.51) are given by the following Theorem.

Theorem 6.4.4. *The normalized gradient algorithm with projection described by the equation (6.74) and designed for the parametric model (6.51) guarantees that*
(i) $\epsilon, \epsilon n_s, \theta, \dot{\theta} \in L_\infty$
(ii) $\epsilon, \epsilon n_s, \dot{\theta} \in \mathcal{S}\left(\frac{\eta^2}{m^2}\right)$
(iii) If $\eta = 0$ *then* $\epsilon, \epsilon n_s, \dot{\theta} \in L_2$.

Proof. As in Chapter 4, we can establish that the projection guarantees that $\theta(t) \in \mathcal{C}\ \forall\, t \geq 0$ provided $\theta(0) \in \mathcal{C}$. Thus $\theta, \tilde{\theta} \in L_\infty$. Since $\epsilon = \frac{-\tilde{\theta}^T\phi+\eta}{m^2}$ and $\frac{\phi}{m}, \frac{\eta}{m} \in L_\infty$, it follows that $\epsilon, \epsilon n_s \in L_\infty$. Furthermore, from (6.74) and the fact that orthogonal projection always decreases the norm of a vector, we have $|\dot{\theta}| \leq \gamma|\epsilon m|\frac{|\phi|}{m}$ which implies that $\dot{\theta} \in L_\infty$. This completes the proof of property (i).

Let us now choose the Lyapunov-like function

$$V = \frac{\tilde{\theta}^T\tilde{\theta}}{2\gamma}$$

Then along the solution of (6.74) we have

$$\dot{V} = \begin{cases} -\epsilon^2 m^2 + \epsilon\eta & \text{if } \theta \in \mathcal{C}^0 \text{ or if } \theta \in \delta\mathcal{C} \text{ and } (\gamma\epsilon\phi)^T\nabla g \leq 0 \\ -\epsilon^2 m^2 + \epsilon\eta - \tilde{\theta}^T\frac{(\nabla g \nabla g^T)}{\nabla g^T \nabla g}\epsilon\phi & \text{otherwise} \end{cases}$$

Now for $\theta \in \delta\mathcal{C}$ and $(\gamma\epsilon\phi)^T\nabla g > 0$, we have

$$\operatorname{sgn}\left\{\frac{\tilde{\theta}^T \nabla g \nabla g^T \epsilon\phi}{\nabla g^T \nabla g}\right\} = \operatorname{sgn}\{\tilde{\theta}^T \nabla g\}.$$

Since $\mathcal{C}$ is convex and $\theta^* \in \mathcal{C}$, we have $\tilde{\theta}^T \nabla g = (\theta - \theta^*)^T \nabla g \geq 0$ when $\theta \in \delta\mathcal{C}$. Therefore, it follows that $\frac{\tilde{\theta}^T \nabla g \nabla g^T \epsilon\phi}{\nabla g^T \nabla g} \geq 0$ when $\theta \in \delta\mathcal{C}$ and $(\gamma\epsilon\phi)^T \nabla g > 0$. Hence, the term due to projection can only make $\dot{V}$ more negative and, therefore, for all $t \geq 0$, we have

$$\dot{V} = -\epsilon^2 m^2 + \epsilon\eta \leq -\frac{\epsilon^2 m^2}{2} + \frac{\eta^2}{2m^2}$$

Since V is bounded due to $\theta \in L_\infty$, it follows that $\epsilon m \in \mathcal{S}\left(\frac{\eta^2}{m^2}\right)$ which implies that $\epsilon, \epsilon n_s \in \mathcal{S}\left(\frac{\eta^2}{m^2}\right)$. Since from (6.74) $|\dot{\theta}| \leq \gamma|\epsilon m|\frac{|\phi|}{m}$ and $\frac{\phi}{m} \in L_\infty$, it follows that $\dot{\theta} \in \mathcal{S}\left(\frac{\eta^2}{m^2}\right)$ and the proof of (ii) is complete. The proof of property (iii) follows by setting $\eta = 0$ and repeating the above arguments.

6.4.4 Dead Zone

Let us consider the normalized estimation error

$$\epsilon = \frac{z - \theta^T\phi}{m^2} = \frac{-\tilde{\theta}^T\phi + \eta}{m^2} \tag{6.75}$$

for the parametric model (6.51). The signal ϵ which is used to "drive" the normalized gradient adaptive law is a measure of the parameter error $\tilde{\theta}$, which is present in the signal $\tilde{\theta}^T\phi$, and of the modelling error η. When $\eta = 0$ and $\tilde{\theta} = 0$ we have $\epsilon = 0$ and no adaptation takes place. Because $\frac{\eta}{m}, \frac{\phi}{m} \in L_\infty$, large ϵm implies that $\frac{\tilde{\theta}^T\phi}{m}$ is large which in turn implies that $\tilde{\theta}$ is large. In this case, the effect of the modelling error η is small and the parameter estimates driven by ϵ move in a direction which reduces $\tilde{\theta}$. On the other hand, when ϵm is small, the effect of η may be more dominant than that of the signal $\tilde{\theta}^T\phi$ and the parameter estimates may be driven in a direction dictated predominantly by η. The principal idea behind the deadzone, therefore, is to monitor the size of the estimation error and adapt only when the estimation error is large relative to the modelling error.

Let us consider the gradient algorithm for the linear parametric model (6.51). We consider the same cost function as in the ideal case, i.e.

$$J(\theta,\, t) = \frac{(z - \theta^T\phi)^2}{2m^2} = \frac{\epsilon^2 m^2}{2}$$

and write

$$\dot{\theta} = \begin{cases} -\gamma\nabla J(\theta) & \text{if } |\epsilon m| > g_0 > \frac{|\eta|}{m} \\ 0 & \text{otherwise} \end{cases} \tag{6.76}$$

In other words, we move in the direction of steepest descent only when the estimation error is large relative to the modelling error i.e. when $|\epsilon m| > g_0$ and switch off adaptation when ϵm is small i.e. $|\epsilon m| \leq g_0$. In view of (6.76) we have

$$\dot{\theta} = \gamma\phi(\epsilon + g), \; g = \begin{cases} 0 & \text{if } |\epsilon m| > g_0 \\ -\epsilon & \text{if } |\epsilon m| \leq g_0 \end{cases} \tag{6.77}$$

To avoid any implementation problems which may arise due to the discontinuity in (6.77) [35], the Dead Zone function is made continuous as follows

$$\dot{\theta} = \gamma\phi(\epsilon + g), \; g = \begin{cases} \frac{g_0}{m} & \text{if } \epsilon m < -g_0 \\ -\frac{g_0}{m} & \text{if } \epsilon m > g_0 \\ -\epsilon & \text{if } |\epsilon m| \leq g_0 \end{cases} \tag{6.78}$$

The continuous and discontinuous dead-zone functions are shown in Fig. 6.4. Because the size of the Dead Zone depends on m, this Dead Zone function is

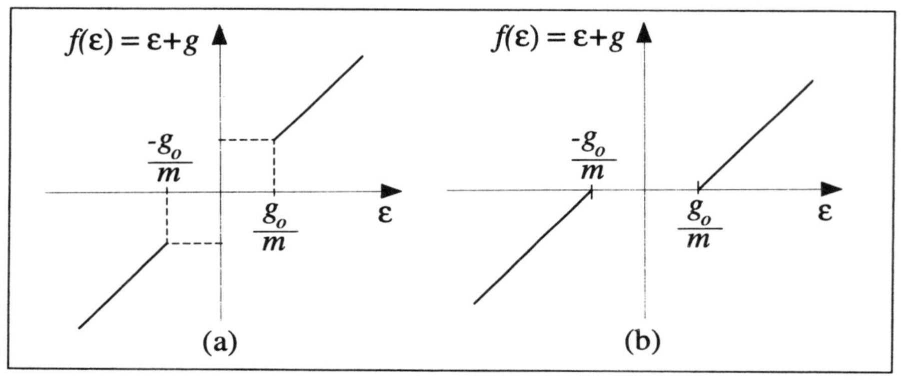

Fig. 6.4. Normalized Dead Zone Functions: (a) discontinuous (b) continuous

often referred to as the normalized Dead Zone.

The stability properties of (6.78) are given by the following Theorem.

Theorem 6.4.5. *The adaptive law (6.78) designed for the parametric model (6.51) guarantees that*

(i) $\epsilon, \epsilon n_s, \theta, \dot{\theta} \in L_\infty$

(ii) $\epsilon, \epsilon n_s \in \mathcal{S}\left(g_0 + \frac{\eta^2}{m^2}\right)$

(iii) $\dot{\theta} \in L_2$

Proof. Consider the Lyapunov-like function $V(\tilde{\theta}) = \frac{\tilde{\theta}^T\tilde{\theta}}{2\gamma}$ whose derivative along the solution of (6.78) is given by

$$\begin{aligned} \dot{V} &= \tilde{\theta}^T\phi(\epsilon + g) \\ &= -(\epsilon m^2 - \eta)(\epsilon + g) \text{ (using (6.75))} \end{aligned}$$

From (6.78), we see that

$$(\epsilon m^2 - \eta)(\epsilon + g) = \begin{cases} (\epsilon m - \frac{\eta}{m})(\epsilon m + g_0) > 0 & \text{if } \epsilon m < -g_0 < -\frac{|\eta|}{m} \\ (\epsilon m - \frac{\eta}{m})(\epsilon m - g_0) > 0 & \text{if } \epsilon m > g_0 > \frac{|\eta|}{m} \\ 0 & \text{if } |\epsilon m| \leq g_0 \end{cases}$$

Thus $(\epsilon m^2 - \eta)(\epsilon + g) \geq 0 \ \forall \ t \geq 0$ so that $\dot{V} \leq 0$ which implies that $V, \tilde{\theta}, \theta \in L_\infty$ and $\sqrt{(\epsilon m^2 - \eta)(\epsilon + g)} \in L_2$. The boundedness of θ, $\frac{\phi}{m}$, $\frac{\eta}{m}$ along with (6.75) implies that $\epsilon, \epsilon n_s \in L_\infty$. This along with (6.78) implies that $\dot{\theta} \in L_\infty$ and the proof of property (i) is complete.

To prove (ii), we note that

$$\begin{aligned} \dot{V} &= -(\epsilon m^2 - \eta)(\epsilon + g) \\ &= -\epsilon^2 m^2 - (\epsilon m)(gm) + \epsilon m \frac{\eta}{m} + gm\frac{\eta}{m} \\ &\leq -\epsilon^2 m^2 + |\epsilon m| g_0 + |\epsilon m| \frac{|\eta|}{m} + g_0 \frac{|\eta|}{m} \text{ (using (6.78))} \\ &\leq -\frac{\epsilon^2 m^2}{2} + \frac{|\eta|^2}{m^2} + g_0^2 + \frac{|\eta|}{m} g_0 \text{ (completing squares twice)} \\ &\leq -\frac{\epsilon^2 m^2}{2} + \frac{3}{2}\frac{|\eta|^2}{m^2} + \frac{3}{2} g_0^2 \text{ (using } xy \leq \tfrac{x^2+y^2}{2}) \end{aligned}$$

which, together with $V \in L_\infty$, implies that $\epsilon m \in \mathcal{S}\left(\frac{\eta^2}{m^2} + g_0\right)$. Since $\epsilon^2 m^2 = \epsilon^2 + \epsilon^2 n_s^2$, it follows that $\epsilon, \epsilon n_s \in \mathcal{S}\left(\frac{\eta^2}{m^2} + g_0\right)$ and the proof of (ii) is complete.

To prove (iii), we note that from (6.78), we have

$$(\epsilon m + gm)^2 = \begin{cases} (\epsilon m + g_0)^2 & \text{if } \epsilon m < -g_0 \\ (\epsilon m - g_0)^2 & \text{if } \epsilon m > g_0 \\ 0 & \text{if } |\epsilon m| \leq g_0 \end{cases}$$

Furthermore,

$$(\epsilon m^2 - \eta)(\epsilon + g) = \begin{cases} (\epsilon m + g_0)^2 - (g_0 + \frac{\eta}{m})(\epsilon m + g_0) & \text{if } \epsilon m < -g_0 \\ (\epsilon m - g_0)^2 + (g_0 - \frac{\eta}{m})(\epsilon m - g_0) & \text{if } \epsilon m > g_0 \\ 0 & \text{if } |\epsilon m| \leq g_0 \end{cases}$$

Thus $(\epsilon m^2 - \eta)(\epsilon + g) \geq (\epsilon m + gm)^2$. Since $\sqrt{(\epsilon m^2 - \eta)(\epsilon + g)} \in L_2$ (see proof of property (i)), it follows that $(\epsilon + g)m \in L_2$. Now from (6.78) $\dot{\theta} = \gamma\phi(\epsilon + g) = \gamma\frac{\phi}{m}(\epsilon + g)m$. Since $\frac{\phi}{m} \in L_\infty$, we conclude that $\dot{\theta} \in L_2$ and the proof of (iii) is complete.

CHAPTER 7
ROBUST ADAPTIVE IMC SCHEMES

7.1 Introduction

In Chapter 5, we considered the design and analysis of adaptive internal model control (AIMC) schemes under ideal conditions, i.e. in the absence of modelling errors. The key idea was to combine an IMC control structure for the known parameter case, presented in Chapter 3, with a parameter estimator from Chapter 4 in a Certainty Equivalence fashion. Although in the absence of modelling errors, such schemes were proven to be stable, the instability examples presented in the last chapter clearly demonstrated that the adaptive laws of Chapter 4 could lead to totally unacceptable behaviour in the presence of modelling errors. The situation is likely to be even worse when such an adaptive law is used to implement a Certainty Equivalence adaptive control scheme, and modelling errors are present. In the last chapter, we also discussed several approaches for correcting this kind of erratic behaviour of the adaptive laws leading to the design of adaptive laws that are robust to the presence of a class of modelling errors. Since the IMC schemes of Chapter 3 were shown to be inherently robust to the presence of modelling errors, a natural question that comes to mind is whether robust adaptive IMC schemes can be designed by combining an IMC controller structure with a *robust* adaptive law. The objective of this chapter is to provide an affirmative answer to this question by showing that whenever any of the robust adaptive laws of the last chapter is combined with any of the IMC controller structures from Chapter 3 in a Certainty Equivalence fashion, the result is a robust adaptive IMC scheme.

For clarity of presentation, in this chapter, we will assume that the plant uncertainty is modelled as a stable multiplicative perturbation. To this end, let the plant be described by the input-output relationship

$$y = \frac{Z_0(s)}{R_0(s)}[1+\mu\Delta_m(s)][u],\ \mu > 0 \tag{7.1}$$

where u, y are the plant input and output signals; $P_0(s) = \frac{Z_0(s)}{R_0(s)}$ represents the transfer function of the modelled part of the plant; $R_0(s)$ is a monic Hurwitz polynomial of degree n; $Z_0(s)$ is a polynomial of degree l with $l < n$;

the coefficients of $Z_0(s)$, $R_0(s)$ are unknown; and $\mu\Delta_m(s)$ is a stable multiplicative uncertainty such that $\frac{Z_0(s)}{R_0(s)}\Delta_m(s)$ is strictly proper. In order to implement a certainty equivalence based robust AIMC scheme, we need to first design a robust parameter estimator to provide on-line estimates of the unknown parameters. This can be done using the results in Chapter 6.

7.2 Design of the Robust Adaptive Law

Let $\Lambda(s)$ be an arbitrary monic Hurwitz polynomial of degree n. Then, as in Chapter 6, the plant equation (7.1) can be rewritten as

$$y = \theta^{*^T} \phi + \eta \tag{7.2}$$

where $\theta^* = [\theta_1^{*^T}, \theta_2^{*^T}]^T$; θ_1^*,θ_2^* are vectors containing the coefficients of $[\Lambda(s) - R_o(s)]$ and $Z_o(s)$ respectively;

$$\begin{aligned} \phi &= [\phi_1^T, \phi_2^T]^T \qquad (7.3) \\ \phi_1 &= \frac{a_{n-1}(s)}{\Lambda(s)}[y],\ \phi_2 = \frac{a_l(s)}{\Lambda(s)}[u] \\ a_{n-1}(s) &= \left[s^{n-1}, s^{n-2}, \ldots, 1\right]^T \\ a_l(s) &= \left[s^l, s^{l-1}, \ldots, 1\right]^T \\ \text{and } \eta &\triangleq \mu\frac{\Delta_m(s)Z_o(s)}{\Lambda(s)}[u]. \qquad (7.4) \end{aligned}$$

Equation (7.2) is exactly in the form of the linear parametric model with modelling error considered in Chapter 6 for which a large class of robust adaptive laws can be developed.

Since the parameter estimates obtained from these robust adaptive laws are to be used for *control design* purposes, we cannot apriori assume the boundedness of the signal vector ϕ or that of η. Thus dynamic normalization as introduced in Chapter 6 will be required. Furthermore, depending on the particular IMC scheme being considered, the "estimated modelled part of the plant" based on the parameter estimates will have to satisfy certain properties: for instance, for an adaptive model reference control scheme, the estimated modelled part of the plant will have to be pointwise minimum-phase. These specific requirements will vary from one controller to the other and will, therefore, be discussed in detail in the next section when we consider the actual control designs. Nevertheless, since the estimated modelled part of the plant will have to satisfy these properties, it is imperative that such a feature be built into the parameter estimator. To do so, we assume that convex sets $\mathcal{C}_\theta$, which will differ from one IMC scheme to the other, are known in the parameter space such that for every θ in $\mathcal{C}_\theta$, the corresponding plant satisfies the desired properties. Then by projecting θ onto $\mathcal{C}_\theta$, we can ensure that the parameter estimates do have the desired properties required for a particular control design. Using the results of Section 6.4 of the last chapter, we can design quite a few different robust adaptive laws. For instance, using the *dynamically normalized gradient* algorithm, with *parameter projection* as the robustifying modification, we obtain the following robust adaptive law:

$$\dot{\theta} = Pr[\gamma\varepsilon\phi],\ \theta(0) \in \mathcal{C}_\theta \tag{7.5}$$

$$\varepsilon = \frac{y - \hat{y}}{m^2} \tag{7.6}$$

$$\hat{y} = \theta^T \phi \tag{7.7}$$

$$m^2 = 1 + n_s^2,\ n_s^2 = m_s \tag{7.8}$$

$$\dot{m}_s = -\delta_o m_s + u^2 + y^2,\ m_s(0) = 0 \tag{7.9}$$

where $\gamma > 0$ is an adaptive gain; $\mathcal{C}_\theta$ is a known compact convex set containing θ^*; $Pr[\cdot]$ is the projection operator given by (6.74) which guarantees that the parameter estimate $\theta(t)$ does not exit the set $\mathcal{C}_\theta$; and $\delta_o > 0$ is a constant chosen so that $\Delta_m(s), \frac{1}{\Lambda(s)}$ are analytic in $\mathcal{R}e[s] \geq -\frac{\delta_o}{2}$. This choice of δ_o, of course, necessitates the following apriori knowledge about the stability margin of the unmodelled dynamics:

A1. $\Delta_m(s)$ is analytic in Re$[s] \geq -\frac{\delta_{01}}{2}$ for some known $\delta_{01} > 0$.

We recall that assumption A1 above, along with the assumption about $\frac{\Delta_m(s)Z_0(s)}{\Lambda(s)}$ being strictly proper, was used in Chapter 6 to achieve a design for the normalizing signal m to ensure that $\frac{\eta}{m} \in L_\infty$. Furthermore, it is important to point out that the compactness assumption on $\mathcal{C}_\theta$ here, in addition to the usual convexity requirement, ensures the boundedness of the parameter estimates. Later on, we will see that the compactness assumption also aids in the stability analysis.

Various robust adaptive IMC schemes can now be obtained by replacing the internal model in Fig. 3.5 by that obtained from Equation (7.7), and the IMC parameters $Q(s)$ by time-varying operators which implement the certainty equivalence versions of the controller structures considered in Chapter 3. The detailed design of these Certainty Equivalence controllers is discussed next.

7.3 Certainty Equivalence Control Laws

We first outline the steps involved in designing a general Certainty Equivalence Robust Adaptive IMC scheme. Thereafter, additional simplifications or complexities that result from the use of a particular control law will be discussed.

- **Step 1:** First use the parameter estimate $\theta(t)$ obtained from the robust adaptive law (7.5)-(7.9) to generate estimates of the numerator and denominator polynomials for the modelled part of the plant[1]:

$$\begin{aligned}\hat{Z}_o(s,t) &= \theta_2^T(t)a_l(s)\\ \hat{R}_o(s,t) &= \Lambda(s) - \theta_1^T(t)a_{n-1}(s)\end{aligned}$$

- **Step 2:** Using the frozen time modelled part of the plant $\hat{P}_0(s,t) = \frac{\hat{Z}_o(s,t)}{\hat{R}_o(s,t)}$, calculate the appropriate $\hat{Q}(s,t)$ using the results developed in Chapter 3.
- **Step 3:** Express $\hat{Q}(s,t)$ as $\hat{Q}(s,t) = \frac{\hat{Q}_n(s,t)}{\hat{Q}_d(s,t)}$ where $\hat{Q}_n(s,t)$ and $\hat{Q}_d(s,t)$ are time-varying polynomials with $\hat{Q}_d(s,t)$ *being monic.*
- **Step 4:** Choose $\Lambda_1(s)$ to be an arbitrary monic Hurwitz polynomial of degree equal to that of $\hat{Q}_d(s,t)$, and let this degree be denoted by n_d.
- **Step 5:** In view of the equivalent IMC representation (3.5) introduced in Chapter 3, the certainty equivalence control law becomes

$$u = q_d^T(t)\frac{a_{n_d-1}(s)}{\Lambda_1(s)}[u] + q_n^T(t)\frac{a_{n_d}(s)}{\Lambda_1(s)}[r - \epsilon m^2] \tag{7.10}$$

where $q_d(t)$ is the vector of coefficients of $\Lambda_1(s)-\hat{Q}_d(s,t)$; $q_n(t)$ is the vector of coefficients of $\hat{Q}_n(s,t)$; $a_{n_d}(s) = [s^{n_d}, s^{n_d-1}, \cdots, 1]^T$ and $a_{n_d-1}(s) = [s^{n_d-1}, s^{n_d-2}, \cdots, 1]^T$.

The robust adaptive IMC scheme resulting from combining the control law (7.10) with the robust adaptive law (7.5)-(7.9) is schematically depicted in Fig. 7.1. We now proceed to discuss the simplifications or additional complexities that result from the use of each of the controller structures presented in Chapter 3.

7.3.1 Robust Adaptive Partial Pole Placement

In this case, the design of the IMC parameter does not depend on the estimated plant. Indeed, $Q(s)$ is a fixed stable transfer function and not a time varying operator. Since the estimated modelled part of the plant is not used for on-line control design purposes, the parameter estimates do not have to satisfy any special properties. As a result, in this case, parameter projection is required in the adaptive law only for guaranteeing the boundedness of the estimated parameters.

---

[1] As in Chapter 5 the "hats" here denote the time varying polynomials/frozen time "transfer functions" that result from replacing the time-invariant coefficients of a "hat-free" polynomial/transfer function corresponding to the modelled part of the plant by their time-varying values obtained from adaptation and/or certainty equivalence control.

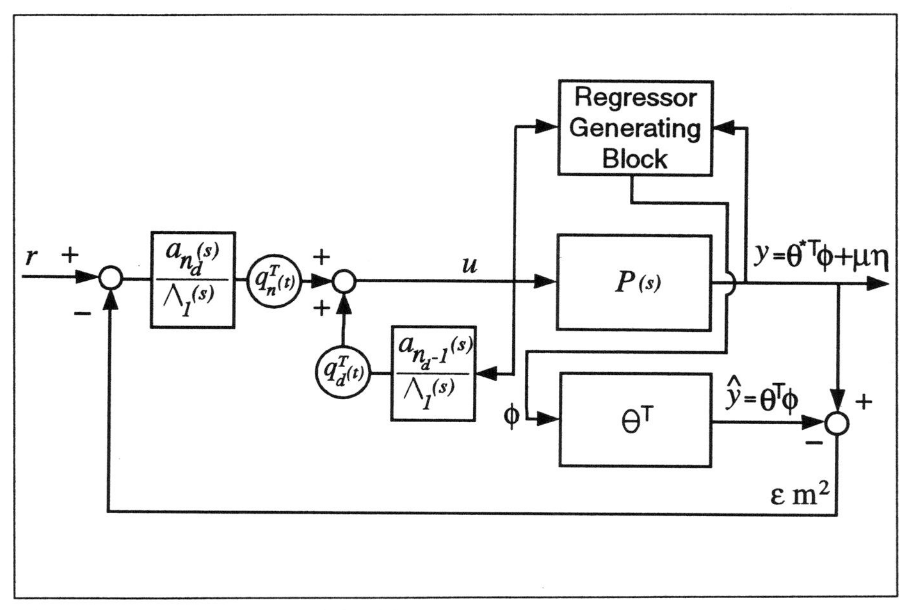

Fig. 7.1. Robust Adaptive IMC Scheme

7.3.2 Robust Adaptive Model Reference Control

In this case from (3.6), we see that the $\hat{Q}(s,t)$ in Step 2 of the Certainty Equivalence design becomes

$$\hat{Q}(s,t) = W_m(s)\left[\hat{P}_0(s,t)\right]^{-1}. \tag{7.11}$$

Our stability and robustness analysis to be presented in Section 7.5 is based on results in the area of slowly time varying systems, in particular Theorem 2.4.9. In order for these results to be applicable, it is required that the operator $\hat{Q}(s,t)$ be pointwise stable (for each fixed t) and also that the degree of $\hat{Q}_d(s,t)$ in Step 3 of the Certainty Equivalence Design not change with time. These two requirements can be satisfied as follows:

- The pointwise stability of $\hat{Q}(s,t)$ can be guaranteed by ensuring that the frozen time estimated modelled part of the plant is minimum phase, i.e. $\hat{Z}_o(s,t)$ is Hurwitz stable for every fixed t. To guarantee such a property for $\hat{Z}_o(s,t)$, the projection set $\mathcal{C}_\theta$ in (7.5)-(7.9) is chosen so that $\forall\, \theta \in \mathcal{C}_\theta$, the corresponding $Z_0(s,\theta) = \theta_2^T a_l(s)$ is Hurwitz stable. By restricting $\mathcal{C}_\theta$ to be a subset of a Cartesian product of closed intervals, results from Parametric Robust Control [23, 1] can be used to construct such a $\mathcal{C}_\theta$. Also, when the projection set $\mathcal{C}_\theta$ cannot be specified as a single convex set, results from *hysteresis switching* using a finite number of convex sets [29] can be used.

- The degree of $\hat{Q}_d(s,t)$ can be rendered time invariant by ensuring that the leading coefficient of $\hat{Z}_o(s,t)$ is not allowed to pass through zero. As shown in Chapter 4, this feature can be built into the adaptive laws, for the ideal case, by assuming some knowledge about the sign and a lower bound on the absolute value of the leading coefficient of $Z_0(s,\theta)$ and then implementing a parameter projection onto a convex set $\mathcal{C}_\theta$. The same approach is equally applicable to any of the robust adaptive laws considered in Chapter 6.

We will, therefore, assume that for IMC based model reference adaptive control, the compact set $\mathcal{C}_\theta$ has been suitably chosen to guarantee that the estimate $\theta(t)$ obtained from (7.5)-(7.9) actually satisfies both of the properties mentioned above.

7.3.3 Robust Adaptive H_2 Optimal Control

In this case, $\hat{Q}(s,t)$ is obtained by replacing $P_M^{-1}(s)$, $B_P^{-1}(s)$ on the right hand side of (3.10) with $\hat{P}_{0M}^{-1}(s,t)$, $\hat{B}_{P_0}^{-1}(s,t)$ where $\hat{P}_{0M}(s,t)$ is the minimum phase portion of $\hat{P}_0(s,t)$ and $\hat{B}_{P_0}(s,t)$ is the Blaschke product containing the open right-half plane zeros of $\hat{Z}_0(s,t)$. Thus $\hat{Q}(s,t)$ is given by

$$\hat{Q}(s,t) = \hat{P}_{0M}^{-1}(s,t)R_M^{-1}(s)[\hat{B}_{P_0}^{-1}(s,t)R_M(s)]_* F(s) \tag{7.12}$$

where $[.]_*$ denotes that after a partial fraction expansion, the terms corresponding to the poles of $\hat{B}_{P_0}^{-1}(s,t)$ are removed, and $F(s)$ is an IMC Filter used to force $\hat{Q}(s,t)$ to be proper. As in Chapter 5, the degree of $\hat{Q}_d(s,t)$ in Step 3 of the Certainty Equivalence Design can be kept constant *using a single fixed* $F(s)$ provided the leading coefficient of $\hat{Z}_o(s,t)$ is not allowed to pass through zero. Additionally $\hat{Z}_0(s,t)$ should not have any zeros on the imaginary axis. A parameter projection, as in the case of model reference adaptive control, can be incorporated into the adaptive law (7.5)-(7.9) to guarantee both of these properties.

7.3.4 Robust Adaptive H_∞ Optimal Control (one interpolation constraint)

In this case, $\hat{Q}(s,t)$ is obtained by replacing $P(s)$, b_1 on the right hand side of (3.13) with $\hat{P}_0(s,t)$, $\hat{b}_1$, i.e.

$$\hat{Q}(s,t) = \left[1 - \frac{W(\hat{b}_1)}{W(s)}\right] \hat{P}_0^{-1}(s,t)F(s) \tag{7.13}$$

where $\hat{b}_1$ is the open right half plane zero of $\hat{Z}_0(s,t)$ and $F(s)$ is the IMC Filter. Since (7.13) assumes the presence of only one open right half plane zero, the estimated polynomial $\hat{Z}_o(s,t)$ must have only one open right half plane zero and none on the imaginary axis. Additionally the leading coefficient of $\hat{Z}_o(s,t)$ should not be allowed to pass through zero so that the degree of $\hat{Q}_d(s,t)$ in Step 3 of the Certainty Equivalence Design can be kept fixed using a single fixed $F(s)$. Once again, both of these properties can be guaranteed by the adaptive law by appropriately choosing the set $\mathcal{C}_\theta$.

The following Theorem describes the stability and performance properties of the robust adaptive IMC schemes presented in this chapter. The proof is rather long and technical and is, therefore, relegated to Section 7.5.

Theorem 7.3.1. *Consider the plant (7.1) subject to the robust adaptive IMC control law (7.5)-(7.9), (7.10), where (7.10) corresponds to any one of the adaptive IMC schemes considered in this chapter and $r(t)$ is a bounded external signal. Then, $\exists\ \mu^* > 0$ such that $\forall\ \mu \in [0, \mu^*)$, all the signals in the closed loop system are uniformly bounded and the error $y - \hat{y} \in S\left(\frac{\eta^2}{m^2}\right)$.*

7.4 Robust Adaptive IMC Design Examples

In this section, we present some examples to illustrate the steps involved in the design of the proposed robust certainty equivalence adaptive controllers.

Example 7.4.1. (Robust Adaptive Partial Pole Placement) We first consider the plant (7.1) with $Z_0(s) = s + 2$, $R_0(s) = s^2 + s + 1$, $\Delta_m(s) = \frac{s+1}{s+3}$ and $\mu = 0.01$. Choosing $\Lambda(s) = s^2 + 2s + 2$, the plant parametrization becomes

$$\begin{aligned} y &= \theta^{*T}\phi + \eta \\ &\text{where} \\ \theta^* &= [1, 1, 1, 2]^T \\ \phi &= \left[\frac{s}{s^2+2s+2}[y], \frac{1}{s^2+2s+2}[y], \frac{s}{s^2+2s+2}[u], \frac{1}{s^2+2s+2}[u]\right]^T \\ \eta &= 0.01\frac{(s+1)(s+2)}{(s+3)(s^2+2s+2)}[u] \end{aligned}$$

Choosing $\mathcal{C}_\theta = [-5.0,\ 5.0] \times [-4.0,\ 4.0] \times [0.1,\ 6.0] \times [-6.0,\ 6.0]$, $\gamma = 1$, $\delta_0 = 0.1$, $Q(s) = \frac{1}{s+4}$ and implementing the adaptive partial pole placement control scheme (7.5)-(7.9), (7.10), with $\theta(0) = [-1.0,\ 2.0,\ 3.0,\ 1.0]^T$ and all other initial conditions set to zero, we obtained the plots shown in Fig. 7.2 for $r(t) = 1.0$ and $r(t) = sin(0.2t)$. From these plots, it is clear that $y(t)$ tracks $\frac{s+2}{(s^2+s+1)(s+4)}[r]$ quite well.

Example 7.4.2. (Robust Adaptive Model Reference Control) Let us now consider the design of an adaptive model reference control scheme for the same plant used in the last example where now the reference model is given by $W_m(s) = \frac{1}{s^2+2s+1}$. The adaptive law (7.5)-(7.9) must now guarantee that the

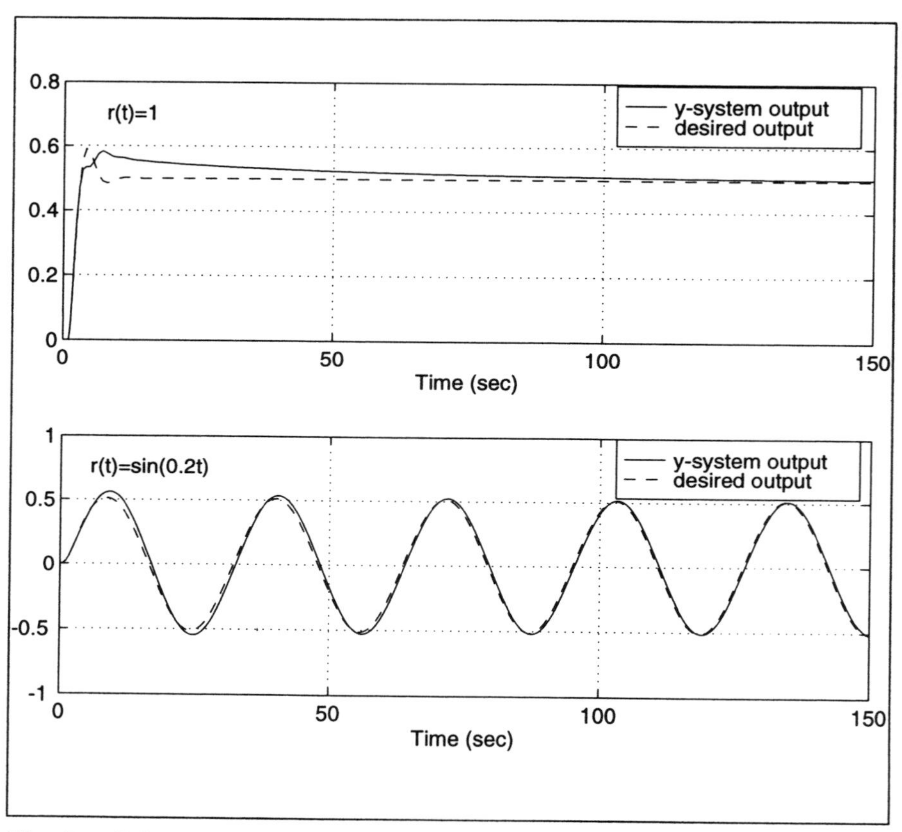

Fig. 7.2. Robust Adaptive Partial Pole Placement Control Simulation

estimated modelled part of the plant is pointwise minimum phase, to ensure which, we now choose the set $\mathcal{C}_\theta$ as $\mathcal{C}_\theta = [-5.0,\ 5.0] \times [-4.0,\ 4.0] \times [0.1,\ 6.0] \times [0.1,\ 6.0]$. All the other design parameters are kept exactly the same as in the last example except that now (7.10) implements the IMC control law (7.11) with $\Lambda_1(s) = s^3 + 2s^2 + 2s + 2$. This choice of a third order $\Lambda_1(s)$ is necessary since from (7.11), it is clear that $n_d = 3$ here. The resulting simulation plots are shown in Fig. 7.3 for $r(t) = 1.0$ and $r(t) = sin(0.2t)$. From these plots, it is clear that the robust adaptive IMC scheme does achieve model following.

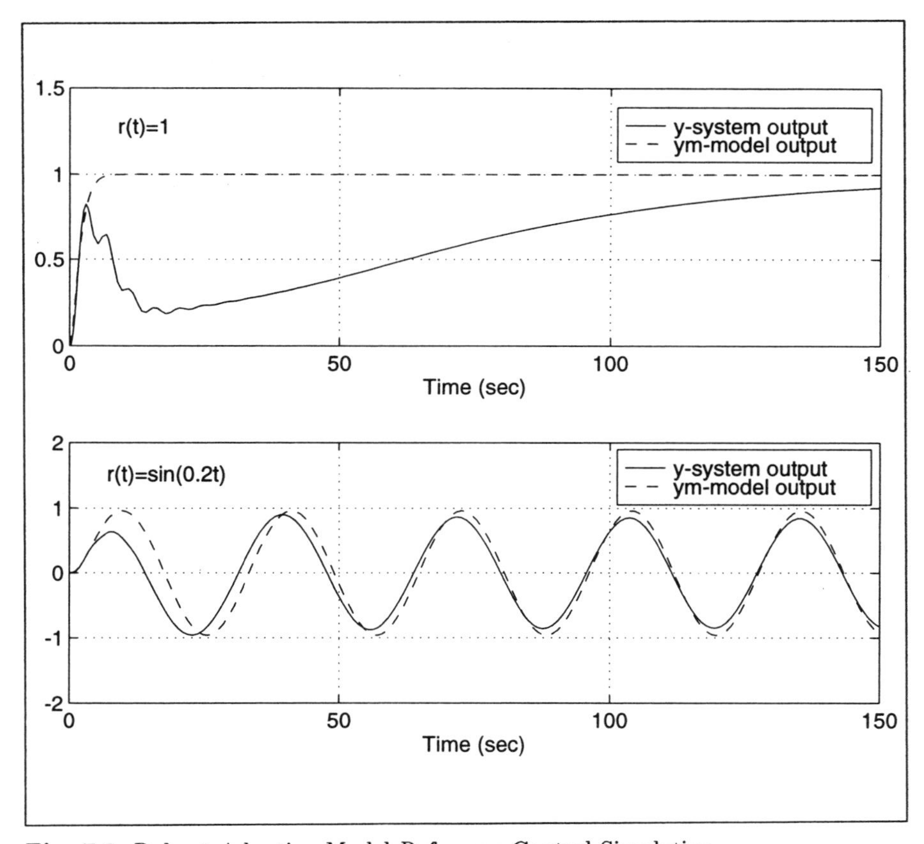

Fig. 7.3. Robust Adaptive Model Reference Control Simulation

Example 7.4.3. (Robust Adaptive H_2 Optimal Control) The modelled part of the plant that we have considered in the last two examples is minimum phase which would not lead to an interesting H_2 or H_∞ optimal control problem. Thus, for H_2 and H_∞ optimal control , we consider the plant (7.1) with $Z_0(s) = -s + 1$, $R_0(s) = s^2 + 3s + 2$, $\Delta_m(s) = \frac{s+1}{s+3}$ and $\mu = 0.01$. Choosing $\Lambda(s) = s^2 + 2s + 2$, the plant parametrization becomes

$$
\begin{aligned}
y &= \theta^{*T}\phi + \eta \\
& \quad \text{where} \\
\theta^* &= [-1, 0, -1, 1]^T \\
\phi &= \left[\frac{s}{s^2+2s+2}[y], \frac{1}{s^2+2s+2}[y], \frac{s}{s^2+2s+2}[u], \frac{1}{s^2+2s+2}[u]\right]^T \\
\eta &= 0.01\frac{(s+1)(-s+1)}{(s+3)(s^2+2s+2)}[u]
\end{aligned}
$$

In order to ensure that $\hat{Z}_0(s,t)$ has no zeros on the imaginary axis and its degree does not drop, the projection set $\mathcal{C}_\theta$ is chosen as $\mathcal{C}_\theta = [-5.0,\ 5.0] \times [-4.0,\ 4.0] \times [-6.0,\ -0.1] \times [0.1,\ 6.0]$. Now choosing $\gamma = 1$, $\delta_0 = 0.1$, $\theta(0) = [-2, 2, -2, 2]^T$, and all other initial conditions equal to zero, we implemented the adaptive law (7.5)-(7.9).

From (7.12), we see that $\hat{Q}(s,t)$ in this case depends on the choice of the command input $r(t)$. Let us choose $r(t)$ to be the unit step function so that $R(s) = R_M(s) = \frac{1}{s}$. Let $\theta(t) = [\theta_1(t), \theta_2(t), \theta_3(t), \theta_4(t)]^T$ denote the estimate of θ^* obtained from the robust adaptive law (7.5)-(7.9). Then the frozen time estimated modelled part of the plant is given by

$$
\hat{P}_0(s,t) = \frac{\theta_3(t)s + \theta_4(t)}{s^2 + (2-\theta_1(t))s + (2-\theta_2(t))}
$$

and its right half plane zero $z_p(t)$ is located at $z_p(t) = -\frac{\theta_4(t)}{\theta_3(t)}$. Furthermore

$$
\hat{P}_0(s,t) = \frac{-s+z_p}{s+\bar{z}_p} \cdot \frac{-\theta_3(s+\bar{z}_p)}{s^2+(2-\theta_1)s+(2-\theta_2)}
$$

so that $\hat{B}_{P_0}(s,t) = \frac{-s+z_p}{s+\bar{z}_p}$ and $\hat{P}_{0_M}(s,t) = \frac{-\theta_3(s+\bar{z}_p)}{s^2+(2-\theta_1)s+(2-\theta_2)}$. Hence, from (7.12), we obtain

$$
\begin{aligned}
\hat{Q}(s,t) &= \frac{s^2+(2-\theta_1)s+(2-\theta_2)}{-\theta_3(s+\bar{z}_p)} . s\left[\frac{s+\bar{z}_p}{-s+z_p} . \frac{1}{s}\right]_* F(s) \\
&= \frac{s^2+(2-\theta_1)s+(2-\theta_2)}{-\theta_3(s+\bar{z}_p)} F(s)
\end{aligned}
$$

It is now clear that to make $\hat{Q}(s,t)$ proper, $F(s)$ must be of relative degree 1. So, let us choose $F(s) = \frac{1}{s+1}$ which results in $n_d = 2$. We now choose $\Lambda_1(s) = s^2 + 2s + 2$, which is of second order, and implement the control law (7.10). The resulting plot is shown in Fig. 7.4. From Fig. 7.4, it is clear that $y(t)$ asymptotically tracks $r(t)$ quite well.

Example 7.4.4. (Robust Adaptive H_∞ Optimal Control) We now design a robust adaptive H_∞ optimal controller for the same plant considered in the last example. The robust adaptive law is exactly the same as the one designed

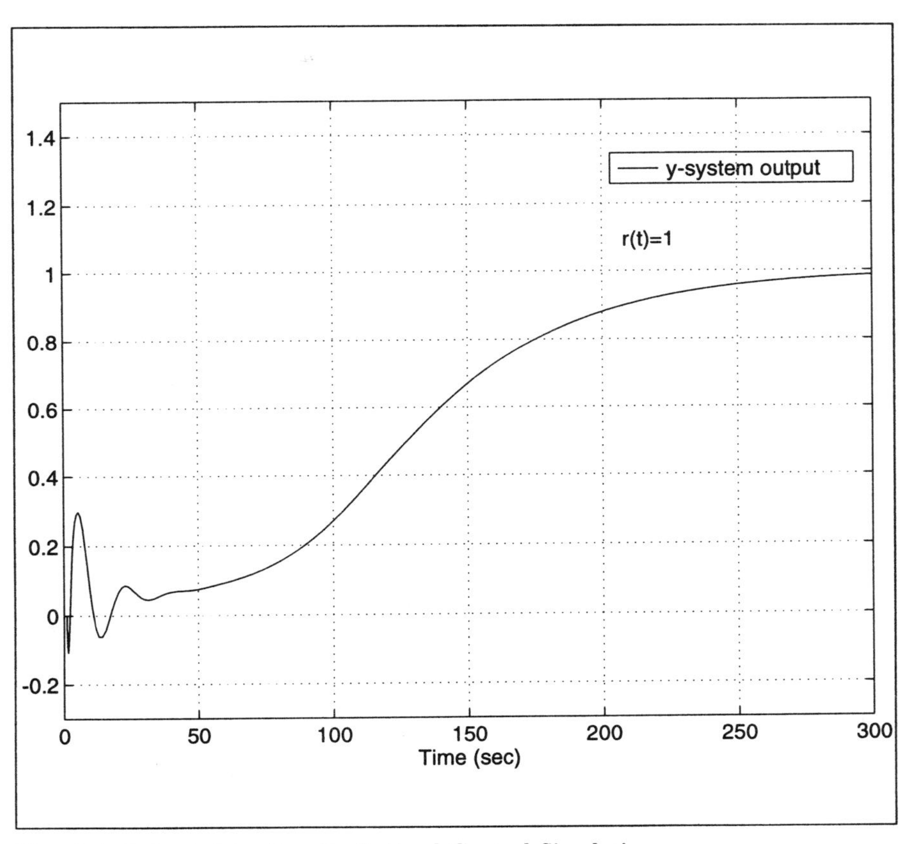

Fig. 7.4. Robust Adaptive H_2 Optimal Control Simulation

earlier and ensures that the frozen-time estimated modelled part of the plant has one and only one right half plane zero. Recall from the previous example that the frozen time estimated modelled part of the plant is given by

$$\hat{P}_0(s,t) = \frac{\theta_3 s + \theta_4}{s^2 + (2-\theta_1)s + (2-\theta_2)}$$

and the right half plane zero is given by $\hat{b}_1 = -\frac{\theta_4}{\theta_3}$.

Choosing $W(s) = \frac{0.01}{s+0.01}$, from (7.13), we have

$$\begin{aligned}\hat{Q}(s,t) &= \left[1 - \frac{s+0.01}{\hat{b}_1 + 0.01}\right] \frac{s^2 + (2-\theta_1)s + (2-\theta_2)}{\theta_3 s + \theta_4} F(s) \\ &= \frac{\hat{b}_1 - s}{\theta_3 s + \theta_4} \frac{(s^2 + (2-\theta_1)s + (2-\theta_2))}{\hat{b}_1 + 0.01} F(s) \\ &= -\frac{1}{\theta_3} \frac{(s^2 + (2-\theta_1)s + (2-\theta_2))}{\hat{b}_1 + 0.01} F(s) \text{ (using } \hat{b}_1 = -\tfrac{\theta_4}{\theta_3})\end{aligned}$$

It is now clear that to make $\hat{Q}(s,t)$ proper, $F(s)$ must be of relative degree 2. So, let us choose $F(s) = \frac{1}{(0.15s+1)^2}$ which results in $n_d = 2$. We now choose $\Lambda_1(s) = s^2+2s+2$ and implement the control law (7.10). Choosing $r(t) = 1.0$ and $r(t) = 0.8sin(0.2t)$, we obtained the plots shown in Fig. 7.5. From these plots, we see that the robust adaptive H_∞-optimal controller does produce reasonably good tracking.

7.5 Stability Proofs of Robust Adaptive IMC Schemes

In this section, we provide the detailed proof of Theorem 7.3.1. The proof makes use of the following two technical lemmas. The proofs of these lemmas are omitted since they have already been presented in Chapter 5 as the proofs of Lemmas 5.5.1 and 5.5.2.

Lemma 7.5.1. *In each of the robust adaptive IMC schemes presented in Section 7.3, the degree of $\hat{Q}_d(s,t)$ in Step 3 of the Certainty Equivalence Design can be made time invariant. Furthermore, for the adaptive H_2 and H_∞ designs, this can be done using a single fixed $F(s)$.*

Lemma 7.5.2. *At any fixed time t, the coefficients of $\hat{Q}_d(s,t)$, $\hat{Q}_n(s,t)$, and hence the vectors $q_d(t)$, $q_n(t)$ are continuous functions of the estimate $\theta(t)$.*

Proof of Theorem 7.3.1: The proof of Theorem 7.3.1 is obtained by combining the properties of the robust adaptive law (7.5)-(7.9) with the properties of the IMC based controller structure. The properties of the robust adaptive law have already been established in Chapter 6. Indeed, since $\frac{\phi}{m}, \frac{\eta}{m} \in L_\infty$,

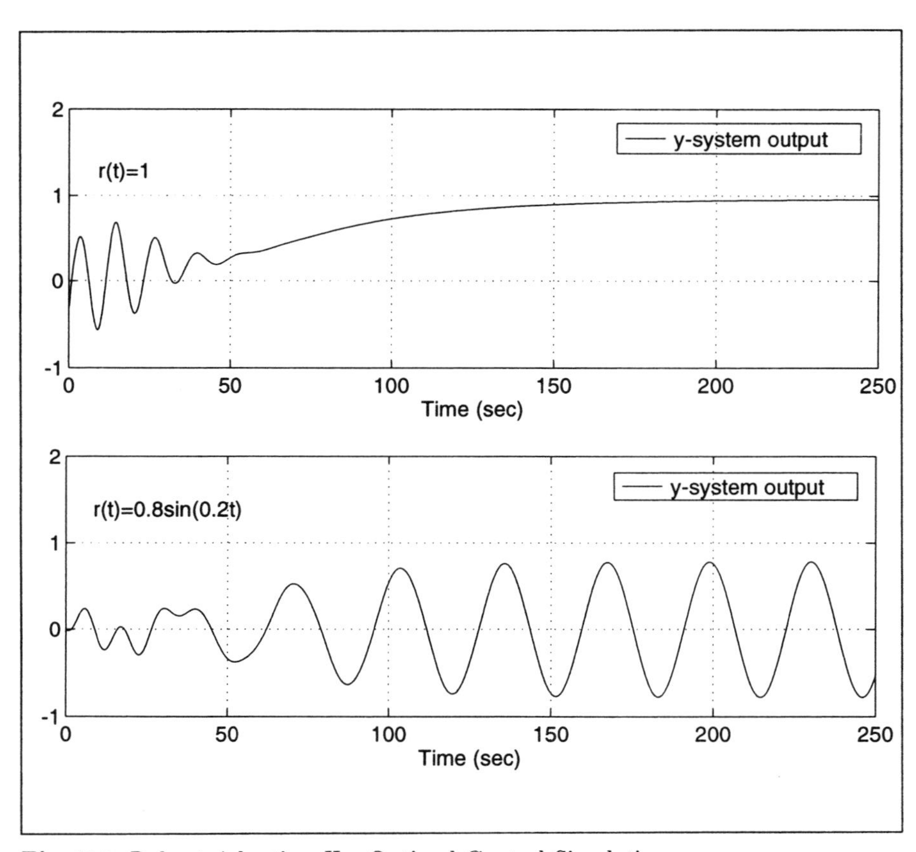

Fig. 7.5. Robust Adaptive H_∞ Optimal Control Simulation

from Theorem 6.4.4, we see that the robust adaptive law (7.5)-(7.9) guarantees that (i) $\epsilon, \epsilon n_s, \theta, \dot{\theta} \in L_\infty$ and (ii) $\epsilon, \epsilon n_s, \dot{\theta} \in \mathcal{S}\left(\frac{\eta^2}{m^2}\right)$. To complete the stability proof, we now turn to the properties of the IMC based controller structure.

The certainty equivalence control law (7.10) can be rewritten as

$$\frac{s^{n_d}}{\Lambda_1(s)}[u] + \beta_1(t)\frac{s^{n_d-1}}{\Lambda_1(s)}[u] + \cdots + \beta_{n_d}(t)\frac{1}{\Lambda_1(s)}[u] = q_n^T(t)\frac{a_{n_d}(s)}{\Lambda_1(s)}[r - \epsilon m^2]$$

where $\beta_1(t)$, $\beta_2(t)$, $\cdots, \beta_{n_d}(t)$ are the time-varying coefficients of $\hat{Q}_d(s,t)$. Defining $x_1 = \frac{1}{\Lambda_1(s)}[u]$, $x_2 = \frac{s}{\Lambda_1(s)}[u]$, $\cdots$,$x_{n_d} = \frac{s^{n_d-1}}{\Lambda_1(s)}[u]$, $X \triangleq [x_1, x_2, \cdots, x_{n_d}]^T$, the above equation can be rewritten as

$$\dot{X} = A(t)X + Bq_n^T(t)\frac{a_{n_d}(s)}{\Lambda_1(s)}[r - \epsilon m^2] \tag{7.14}$$

where

$$A(t) \triangleq \begin{bmatrix} 0 & 1 & 0 & \cdot & \cdot & 0 \\ 0 & 0 & 1 & 0 & \cdot & 0 \\ \cdot & \cdot & \cdot & \cdot & \cdot & \cdot \\ \cdot & \cdot & \cdot & \cdot & \cdot & \cdot \\ -\beta_{n_d}(t) & -\beta_{n_d-1}(t) & \cdot & \cdot & \cdot & -\beta_1(t) \end{bmatrix}$$

$$B \triangleq \begin{bmatrix} 0 \\ 0 \\ \cdot \\ 0 \\ 1 \end{bmatrix}$$

Since the time-varying polynomial $\hat{Q}_d(s,t)$ is pointwise Hurwitz, it follows that for any *fixed* t, the eigenvalues of $A(t)$ are in the open left half plane. Moreover, since the coefficients of $\hat{Q}_d(s,t)$ are continuous functions of $\theta(t)$ (Lemma 7.5.2) and $\theta(t) \in \mathcal{C}_\theta$, a compact set, it follows that $\exists\ \sigma_s > 0$ such that

$$\text{Re}\{\lambda_i(A(t))\} \leq -\sigma_s\ \forall\, t \geq 0 \text{ and } i = 1, 2, \cdots, n_d.$$

The continuity of the elements of $A(t)$ with respect to $\theta(t)$ and the fact that $\dot{\theta} \in \mathcal{S}\left(\frac{\eta^2}{m^2}\right)$ together imply that $\dot{A}(t) \in \mathcal{S}\left(\frac{\eta^2}{m^2}\right)$. From (7.4), (7.8), (7.9), using Lemma 2.3.3(ii), we obtain[2] $\frac{|\eta|}{m} \leq \mu c$. Hence, it follows from Theorem 2.4.9(ii) that $\exists\ \mu_1^* > 0$ such that $\forall\, \mu \in [0, \mu_1^*)$, the equilibrium state $x_e = 0$ of $\dot{x} = A(t)x$ is exponentially stable, i.e. there exist $c_o, p_o > 0$ such that

[2] In the rest of this proof, 'c' is the generic symbol for a positive constant. The exact value of such a constant can be determined (for a quantitative robustness result) as in [42, 15]. However, for the qualitative presentation here, the exact values of these constants are not important.

the state transition matrix $\Phi(t,\tau)$ corresponding to the homogeneous part of (7.14) satisfies

$$\|\Phi(t,\tau)\| \leq c_o e^{-p_o(t-\tau)} \; \forall \; t \geq \tau \tag{7.15}$$

From the identity $u = \frac{\Lambda_1(s)}{\Lambda_1(s)}[u]$, it is easy to see that the control input u can be rewritten as

$$u = v^T(t)X + q_n^T(t)\frac{a_{n_d}(s)}{\Lambda_1(s)}[r - \epsilon m^2] \tag{7.16}$$

$$\begin{aligned} \text{where } v(t) &= [\lambda_{n_d} - \beta_{n_d}(t), \lambda_{n_d-1} - \beta_{n_d-1}(t), \cdots, \lambda_1 - \beta_1(t)]^T \\ \text{and } \Lambda_1(s) &= s^{n_d} + \lambda_1 s^{n_d-1} + \cdots + \lambda_{n_d}. \end{aligned}$$

Also, using (7.16) in the plant equation (7.1), we obtain

$$y = \frac{Z_o(s)}{R_o(s)}[1 + \mu\Delta_m(s)]\left[v^T(t)X + q_n^T(t)\frac{a_{n_d}(s)}{\Lambda_1(s)}[r - \epsilon m^2]\right] \tag{7.17}$$

Now let $\delta \in (0, \min[\delta_o, p_o])$ be chosen such that $\frac{1}{R_o(s)}, \frac{1}{\Lambda_1(s)}$ are analytic in $\mathcal{R}e[s] \geq -\frac{\delta}{2}$, and define the fictitious normalizing signal $m_f(t)$ by

$$m_f(t) = 1.0 + \|u_t\|_2^\delta + \|y_t\|_2^\delta \tag{7.18}$$

We now take truncated exponentially weighted norms on both sides of (7.16),(7.17) and make use of Lemma 2.3.4 and Lemma 2.3.3(i), while observing that $v(t)$,$q_n(t)$,$r(t) \in L_\infty$, to obtain

$$\|u_t\|_2^\delta \leq c + c\|(\varepsilon m^2)_t\|_2^\delta \tag{7.19}$$

$$\|y_t\|_2^\delta \leq c + c\|(\varepsilon m^2)_t\|_2^\delta \tag{7.20}$$

which together with (7.18) imply that

$$m_f(t) \leq c + c\|(\varepsilon m^2)_t\|_2^\delta \tag{7.21}$$

Now squaring both sides of (7.21) we obtain

$$\begin{aligned} m_f^2(t) &\leq c + c\int_0^t e^{-\delta(t-\tau)}\varepsilon^2 m^2 m_f^2(\tau)d\tau \;\; (\text{since } m(t) \leq m_f(t)\;) \\ \Rightarrow m_f^2(t) &\leq c + c\int_0^t e^{-\delta(t-s)}\varepsilon^2(s)m^2(s)\left(e^{c\int_s^t \varepsilon^2 m^2 d\tau}\right)ds \end{aligned}$$

(using Lemma 2.3.5, i.e. the Bellman-Gronwall Lemma)

Since $\varepsilon m \in S\left(\frac{\eta^2}{m^2}\right)$ and $\frac{|\eta|}{m} \leq \mu c$, it follows using Lemma 2.3.2 that $\exists \; \mu^* \in (0, \; \mu_1^*)$ such that $\forall \; \mu \in [0, \mu^*)$, $m_f \in L_\infty$, which in turn implies that $m \in L_\infty$. Since $\frac{\phi}{m}, \frac{\eta}{m}$ are bounded, it follows that $\phi, \; \eta \in L_\infty$. Now, from (7.2), (7.6), (7.7), $\epsilon m^2 = -\tilde{\theta}^T\phi + \eta$ where $\tilde{\theta} \triangleq \theta - \theta^*$. Since $\tilde{\theta} \in L_\infty$, it follows that

$\varepsilon m^2 = -\tilde{\theta}^T \phi + \mu\eta$ is also bounded so that from (7.14), we obtain $X \in L_\infty$. From (7.16), (7.17), we can now conclude that u, $y \in L_\infty$. This establishes the boundedness of all the closed loop signals in the robust adaptive IMC Scheme. Since $y - \hat{y} = \varepsilon m^2$ and $\varepsilon m \in S\left(\frac{\eta^2}{m^2}\right)$, $m \in L_\infty$, it follows that $y - \hat{y} \in S\left(\frac{\eta^2}{m^2}\right)$ as claimed and, therefore, the proof is complete.

Remark 7.5.1. Although the design and analysis in this Chapter was carried out using a specific robust adaptive law, namely the dynamically normalized gradient with projection, it should be clear from the stability analysis presented here that any of the other robust adaptive laws presented in Section 6.4 would have worked just as well.

Remark 7.5.2. In this chapter, for clarity of presentation, we focussed attention on only one type of modelling error, namely a multiplicative plant perturbation. As discussed in [19], other kinds of plant perturbations such as those of the additive or stable-factor type can be handled just as well. Bounded disturbances can also be handled within the same framework although their treatment requires a stability proof by contradiction as opposed to the direct one presented here. The interested reader is referred to [19] for the technical details.

Remark 7.5.3. The robust adaptive IMC scheme of this chapter recovers the performance properties of the ideal case if the modelling error disappears, i.e. we can show that if $\mu = 0$ then $y - \hat{y} \to 0$ as $t \to \infty$. This can be established by incorporating Theorem 6.4.4, Property (iii) into the above analysis and using the same arguments as used for the ideal case considered in Chapter 5. In view of Theorem 6.4.2, property (iii), it follows that a similar result can also be obtained for a robust adaptive law using the switching-σ modification.

CHAPTER 8
CONCLUSION

8.1 Summary

In this research monograph, we have presented a general systematic approach for the design and analysis of robust adaptive IMC schemes for single-input single-output, stable, linear, time-invariant systems. The approach hinges on invoking the ubiquitous Certainty Equivalence Principle [14, 19] of adaptive control to combine robust parameter estimators with robust internal model controller structures to obtain adaptive internal model control schemes that can be proven to be robustly stable. The strategy is very general and can be used by the reader to design and analyze adaptive versions of his or her own favourite IMC scheme. Nevertheless, to get a more concrete feel for the design and analysis procedure, some specific adaptive IMC schemes have been considered, namely those of the partial pole placement, model reference, H_2 optimal and H_∞ optimal control types. The design procedure in each of these cases was illustrated via simple examples.

The principal reason for undertaking the study reported here was the immense popularity of IMC in industrial applications, on the one hand, coupled with the provable theoretical guarantees provided by adaptive control research during the last two decades, on the other. An immediate consequence of this work is to provide a suitable theoretical basis for analytically justifying some of the reported industrial successes of adaptive IMC schemes. In addition, the results of this monograph should facilitate the design of new industrial adaptive IMC controllers with provable guarantees of stability, robustness, performance, etc. It is our hope that this monograph will familiarize adaptive control theorists with the IMC structure while simultaneously arousing the interest of process control engineers in the extremely rich area of adaptive system theory. In that sense, this monograph is like a bridge between the two traditionally diverse areas of adaptive system theory and industrial process control.

8.2 Directions for Future Research

The results presented in this monograph for the design and analysis of adaptive IMC schemes for single-input single-output stable plants are reasonably complete. However, as is quite often the case, they lead us to several other issues that merit further investigation.

First, the results developed here are for the unconstrained single-input single-output case. Since many process control applications are of the constrained multi-input multi-output type, the question that naturally comes to mind is how far the results reported here can be extended to include the more realistic scenario. Given the fact that substantial advances have already been made in the robust adaptive control of multi-input multi-output plants [33, 41], developing similar results for the IMC case may not pose too much of a challenge. However, the introduction of constraints on the magnitudes of the process variables is likely to considerably compound the design and analysis, thereby leading to a challenging research problem.

Second, the robust adaptive IMCs designed in this monograph can tolerate modelling errors such as nonlinearities, time delays, etc. provided they are sufficiently small. However, when the controlled plant in IMC is predominantly nonlinear or when the time delay in the controlled plant happens to be pretty large, the use of a linear plant model in the IMC structure may no longer be appropriate. Instead, the linear plant model in the IMC structure may have to be replaced with a dynamical system containing a function approximator such as a neural network [34]. Of course, further research is necessary to determine what stability properties, if any, could be guaranteed using such a configuration.

Finally, the adaptive IMC results in this monograph have been derived for continuous-time plants. However, the widespread use of digital computers for controls applications makes it appropriate to design and analyze the analogous discrete-time schemes, and perhaps even schemes of the hybrid type. In view of the results presented in [5, 19], such an extension may not be difficult although the precise details would have to be worked out.

From the above discussion, it is clear that each of these issues is important from practical considerations and simultaneously poses interesting theoretical challenges. It is our belief that overcoming some of these challenges will further narrow down the gap between the theory and industrial applications of adaptive control.

APPENDIX A
THE YJBK PARAMETRIZATION OF ALL STABILIZING CONTROLLERS

One of the cornerstones of the modern optimal control theory is the Youla-Jabr-Bongiorno-Kucera (YJBK) parametrization of all stabilizing controllers. Given a linear time invariant finite dimensional plant, described by the rational transfer function $P(s)$, the YJBK parametrization provides a complete characterization of *all* possible linear controllers that stabilize $P(s)$. Such a characterization is very useful for optimization purposes since it facilitates a search over the set of *stabilizing* controllers. It is not our intention here to present a detailed development of the YJBK parametrization since such a detailed presentation can be found in any standard textbook on Robust Control e.g. [43]. Instead, we would like to present the bare minimum necessary to enable the reader to gain an appreciation of the fact that the IMC controller structure is really a special case of the YJBK parametrization.

To this end, we first introduce the notion of internal stability by considering the feedback system shown in Figure A.1. In Figure A.1, $U_1(s)$ and $U_2(s)$

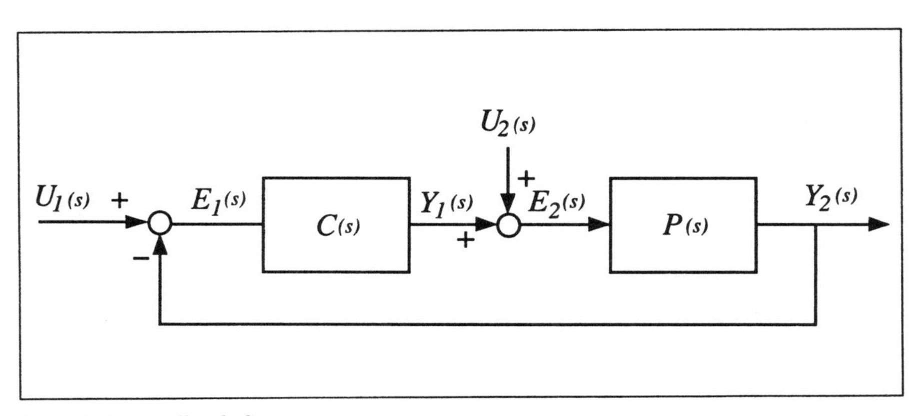

Fig. A.1. Feedback System

are the external inputs; $U_1(s)$ is the exogeneous command signal while $U_2(s)$ is a disturbance input to the plant; $Y_1(s)$ and $Y_2(s)$ are the outputs of the controller $C(s)$ and the plant $P(s)$ respectively while $E_1(s)$ and $E_2(s)$ are the errors produced by the two comparators as shown. From Figure A.1, we obtain

$$\begin{bmatrix} E_1(s) \\ E_2(s) \end{bmatrix} = \begin{bmatrix} U_1(s) \\ U_2(s) \end{bmatrix} - \begin{bmatrix} 0 & P(s) \\ -C(s) & 0 \end{bmatrix} \begin{bmatrix} E_1(s) \\ E_2(s) \end{bmatrix}$$

$$\Rightarrow \left\{ I + \begin{bmatrix} 0 & P(s) \\ -C(s) & 0 \end{bmatrix} \right\} \begin{bmatrix} E_1(s) \\ E_2(s) \end{bmatrix} = \begin{bmatrix} U_1(s) \\ U_2(s) \end{bmatrix}$$

$$\Rightarrow \begin{bmatrix} E_1(s) \\ E_2(s) \end{bmatrix} = \begin{bmatrix} 1/(1+P(s)C(s)) & -P(s)/(1+P(s)C(s)) \\ C(s)/(1+P(s)C(s)) & 1/(1+P(s)C(s)) \end{bmatrix} \cdot \begin{bmatrix} U_1(s) \\ U_2(s) \end{bmatrix}.$$

Define

$$H(P(s), C(s)) = \begin{bmatrix} 1/(1+P(s)C(s)) & -P(s)/(1+P(s)C(s)) \\ C(s)/(1+P(s)C(s)) & 1/(1+P(s)C(s)) \end{bmatrix}.$$

Definition A.0.1. *We say that the controller $C(s)$ stabilizes the plant $P(s)$ if and only if each of the four entries of $H(P(s), C(s))$ is a stable transfer function.*

Note that the present notion of stability is more complete than the usual stability definition given in undergraduate text books where $C(s)$ is said to stabilize $P(s)$ if $P(s)/(1 + P(s)C(s))$ is a stable transfer function. The following example illustrates this fact.

Example A.0.1. Let $P(s) = (s-1)/(s+1)$, $C(s) = 1/(s-1)$. Then $P(s)/(1 + P(s)C(s)) = (s-1)/(s+2)$ so that $C(s)$ stabilizes $P(s)$ as per the undergraduate stability definition. However, one of the entries of $H(P(s), C(s))$ is unstable since $C(s)/(1+P(s)C(s)) = (s+1)/((s+2)(s-1))$ and so $C(s)$ does not stabilize $P(s)$ in the sense of Definition A.0.1

In the following development, given a rational transfer function $P(s)$, we will express it as the ratio of two *stable* rational transfer functions that are *coprime*. However, before doing so, we need to introduce the notion of coprimeness of two stable rational transfer functions.

Definition A.0.2. *Two stable rational transfer functions $N(s)$ and $D(s)$ are said to be coprime if and only if there exist stable rational transfer functions $X(s)$ and $Y(s)$ such that*

$$X(s)N(s) + Y(s)D(s) = 1 \tag{A.1}$$

Equation (A.1) above is referred to as the Bezout identity *or the* Diophantine equation.

We are now ready to state and prove the following lemma on feedback stabilization.

Lemma A.0.1. *Suppose $P(s)$, $C(s)$ are rational transfer functions, and let $P(s) = N_p(s)/D_p(s)$, $C(s) = N_c(s)/D_c(s)$ where $N_p(s)$, $D_p(s)$, $N_c(s)$, $D_c(s)$ are stable rational transfer functions with $N_p(s)$, $D_p(s)$ coprime and $N_c(s)$, $D_c(s)$ coprime, that is, there exist stable rational transfer functions $X_p(s)$, $Y_p(s)$, $X_c(s)$, $Y_c(s)$ such that*

$$\begin{aligned} X_p(s)N_p(s) + Y_p(s)D_p(s) &= 1 \qquad \text{(A.2)} \\ \text{and } X_c(s)N_c(s) + Y_c(s)D_c(s) &= 1. \qquad \text{(A.3)} \end{aligned}$$

Define $\Delta(P(s), C(s)) = N_p(s)N_c(s) + D_p(s)D_c(s)$. Then $C(s)$ stabilizes $P(s)$ if and only if $\Delta(P(s), C(s))$ is minimum phase, that is $1/(\Delta(P(s), C(s)))$ is stable.

Proof. ($\Leftarrow$) Suppose $\Delta(P(s), C(s))$ is minimum phase. Then $1/(\Delta(P(s), C(s)))$ is stable. But

$$H(P(s), C(s)) = 1/(\Delta(P(s), C(s))) \begin{bmatrix} D_p(s)D_c(s) & -N_p(s)D_c(s) \\ D_p(s)N_c(s) & D_p(s)D_c(s) \end{bmatrix} \qquad \text{(A.4)}$$

Thus all the entries of $H(P(s), C(s))$ are stable which implies that $C(s)$ stabilizes $P(s)$.
($\Rightarrow$) Now, suppose that $C(s)$ stabilizes $P(s)$. Then certainly $1 + P(s)C(s) \neq 0$. Also $D_p(s) \neq 0$, $D_c(s) \neq 0$ since both are the denominators of fractions. Hence $\Delta(P(s), C(s)) = D_p(s)D_c(s)(1 + P(s)C(s)) \neq 0$ and the formula (A.4) is valid. Since $H(P(s), C(s))$ has stable entries, it follows from (A.4) that $D_p(s)D_c(s)/(\Delta(P(s), C(s)))$, $D_p(s)N_c(s)/(\Delta(P(s), C(s)))$, $N_p(s)D_c(s)/(\Delta(P(s), C(s)))$ are all stable. Also $1 - D_p(s)D_c(s)/(\Delta(P(s), C(s))) = N_p(s)N_c(s)/(\Delta(P(s), C(s)))$ is stable. A compact way of expressing the stability of the four transfer functions considered above is to say that

$$\begin{bmatrix} D_p(s) \\ N_p(s) \end{bmatrix} 1/(\Delta(P(s), C(s))) \begin{bmatrix} D_c(s) & N_c(s) \end{bmatrix}$$

is stable. Multiplying on the left by $\begin{bmatrix} Y_p(s) & X_p(s) \end{bmatrix}$ and on the right by $\begin{bmatrix} Y_c(s) & X_c(s) \end{bmatrix}^T$ where $Y_p(s)$, $X_p(s)$, $Y_c(s)$, $X_c(s)$ satisfy (A.2) and (A.3), we obtain $1/(\Delta(P(s), C(s)))$ is stable and this completes the proof.

The above lemma leads us to the following Corollary.

Corollary A.0.1. *Suppose $P(s) = N_p(s)/D_p(s)$ where $N_p(s)$, $D_p(s)$ are coprime stable rational transfer functions. Then $C(s)$ stabilizes $P(s)$ if and only if $C(s) = N_c(s)/D_c(s)$ for some stable rational transfer functions $N_c(s)$, $D_c(s)$ that satisfy the Bezout identity*

$$N_p(s)N_c(s) + D_p(s)D_c(s) = 1 \qquad \text{(A.5)}$$

Proof. ($\Leftarrow$) If (A.5) holds then $N_c(s)$, $D_c(s)$ are coprime. Furthermore, $\Delta(P(s), C(s)) = 1$ which is certainly minimum phase. Hence, by Lemma A.0.1, it follows that $C(s)$ stabilizes $P(s)$.
($\Rightarrow$) Now suppose $C(s)$ stabilizes $P(s)$ and express $C(s)$ as $N_1(s)/D_1(s)$ where $N_1(s)$, $D_1(s)$ are stable rational coprime transfer functions. Then by Lemma A.0.1, it follows that $\Delta(P(s), C(s)) = N_1(s)N_p(s) + D_1(s)D_p(s)$ has a stable inverse so that

$$1 = (N_1(s)/(\Delta(P(s), C(s))))N_p(s) + (D_1(s)/(\Delta(P(s), C(s))))D_p(s).$$

Defining $N_c(s) = N_1(s)/(\Delta(P(s), C(s)))$, $D_c(s) = D_1(s)/(\Delta(P(s), C(s)))$, the desired result follows.

We now come to the main result on the YJBK Parametrization of all stabilizing controllers.

Theorem A.0.1. *Suppose $P(s)$ is the rational transfer function of a linear time-invariant plant and let $P(s) = N_p(s)/D_p(s)$ where $N_p(s)$, $D_p(s)$ are coprime stable rational transfer functions. Let $X(s)$ and $Y(s)$ be stable rational transfer functions selected such that*

$$X(s)N_p(s) + Y(s)D_p(s) = 1. \tag{A.6}$$

Then the set of all controllers $C(s)$ that stabilize $P(s)$, denoted by $S(P(s))$ is given by

$$S(P(s)) = \{C(s) = (X(s) + Q(s)D_p(s))/(Y(s) - Q(s)N_p(s)) : Q(s) \textit{ is a stable rational transfer function and } Y(s) - Q(s)N_p(s) \neq 0\}$$

Proof. Suppose that $C(s)$ is of the form $C(s) = (X(s) + Q(s)D_p(s))/(Y(s) - Q(s)N_p(s))$ for some stable rational transfer function $Q(s)$. Then, in view of Corollary A.0.1 and (A.6), $C(s)$ stabilizes $P(s)$ since

$$[X(s) + Q(s)D_p(s)]N_p(s) + [Y(s) - Q(s)N_p(s)]D_p(s) = 1.$$

Conversely, suppose that $C(s)$ stabilizes $P(s)$. Then, from Corollary A.0.1, $C(s) = N_c(s)/D_c(s)$ where $N_c(s)$, $D_c(s)$ are stable rational transfer functions which satisfy

$$N_p(s)N_c(s) + D_p(s)D_c(s) = 1. \tag{A.7}$$

Thus the proof is complete if it can be shown that every solution of (A.7) must be of the form $N_c(s) = X(s) + Q(s)D_p(s)$, $D_c(s) = Y(s) - Q(s)N_p(s)$ for some stable rational transfer function $Q(s)$. Subtracting (A.6) from (A.7) and rearranging terms gives $(N_c(s) - X(s))N_p(s) = (Y(s) - D_c(s))D_p(s)$. Since $N_p(s)$, $D_p(s)$ are coprime, it can be shown (see Appendix A of [43]) that $D_p(s)$ divides $(N_c(s) - X(s)$ and $N_p(s)$ divides $(Y(s) - D_c(s))$ i.e. the quotient $Q(s) = (N_c(s) - X(s))/D_p(s)$ is also a stable rational transfer function. Thus $N_c(s) = X(s) + D_p(s)Q(s)$ and $D_c(s) = Y(s) - Q(s)N_p(s)$ where $Q(s)$ is the above quotient which is itself a stable rational transfer function. This completes the proof.

APPENDIX B
OPTIMIZATION USING THE GRADIENT METHOD

The gradient method or the steepest descent method is a procedure that can be used to iteratively minimize a scalar valued cost function of a vector. The steepest descent method and its variations enjoy immense popularity in the optimization literature. It is not our intention here to present a detailed development of the steepest descent method and its variations. Such a treatment is available in several texts such as [27]. Instead, our objective here is to develop the steepest descent method by intuitively justifying why it can be used as a procedure for iterative minimization of a cost function.

Consider the scalar valued cost function $J(\theta)$ $J : R^n \mapsto R$, where the vector $\theta \in R^n$. Let θ_1, θ_2, $\cdots, \theta_n$ be the individual components of θ. Then the gradient of J with respect to θ, denoted by $\nabla J(\theta)$ is the column vector $[\partial J/\partial\theta_1, \partial J/\partial\theta_2, \cdots \partial J/\partial\theta_n]^T$. Suppose that we are interested in choosing θ iteratively so that the value of the cost function $J(\theta)$ decreases progressively, or at least does not increase, at each step. Let $\theta(k)$ be the value of θ at the kth step. Then, assuming that $\theta(k+1)$ is sufficiently close to $\theta(k)$, we can use the Taylor's series expansion to write

$$J(\theta(k+1)) \approx J(\theta(k)) + \nabla^T J(\theta(k))[\theta(k+1) - \theta(k)] \tag{B.1}$$

where the higher order terms in the Taylor's series expansion have been dropped because of the assumed proximity of $\theta(k+1)$ to $\theta(k)$. Now, from the Cauchy-Schwartz inequality [27], it follows that

$$|\nabla^T J(\theta(k))[\theta(k+1) - \theta(k)]| \leq |\nabla J(\theta(k))||\theta(k+1) - \theta(k)|$$

and furthermore, equality results when $\theta(k+1) - \theta(k) = \alpha[\nabla J(\theta(k))]$ for some $\alpha \in R$, that is $\theta(k+1) - \theta(k)$ is aligned with $\nabla J(\theta(k))$. Thus for a given magnitude of $[\theta(k+1) - \theta(k)]$, $|\nabla^T J(\theta(k))[\theta(k+1) - \theta(k)]|$ will be maximized when $\theta(k+1) - \theta(k)$ and $\nabla J(\theta(k))$ are multiples of each other. Furthermore, it follows that when $\theta(k+1) - \theta(k)$ is in the direction of $-\nabla J(\theta(k))$, i.e. the negative gradient, the decrease in $J(\theta)$ in going from $\theta(k)$ to $\theta(k+1)$ will be maximum. However, one must keep in mind that the above arguments are valid only locally since (B.1) is no longer valid when $\theta(k+1)$ is far away from $\theta(k)$. Nevertheless, for small step sizes, the negative gradient $-\nabla J(\theta(k))$ is the direction of steepest descent and can be used for iteratively decreasing the value of the cost function. This leads us to the steepest descent algorithm

$$\theta(k+1) = \theta(k) - \lambda \nabla J(\theta(k)) \tag{B.2}$$

where $\lambda > 0$ is the step size. Usually, the optimum value of λ, which varies from step to step, is determined by a linear search procedure to minimize the value of the cost function in the direction of steepest descent. However, in this monograph, we are concerned only with the continuous time version of (B.2) which can be derived as follows.

Taking the limit in (B.2) as the step size becomes infinitesimally small, i.e. as λ approaches zero, we obtain

$$\dot{\theta} = -\nabla J(\theta). \tag{B.3}$$

As shown in [19], the gradient descent direction can be scaled using any positive definite symmetric matrix Γ without destroying its properties. As a special case, we can use the scaling $\Gamma = \gamma I$, $\gamma > 0$ resulting in

$$\dot{\theta} = -\gamma \nabla J(\theta), \gamma > 0 \tag{B.4}$$

which is the gradient descent algorithm used for deriving the adaptive laws in this monograph.

In some optimization problems, one may have some apriori knowledge about the whereabouts of the optimum value of θ such as θ belonging to some pre-specified set in R^n. In the case where the optimum value is known to belong to a convex set $\mathcal{C}$, one can modify the gradient descent algorithm (B.4) by using the so called *Gradient Projection Method.* The key idea behind this method is to initiate the search from a value of θ within the convex set and use the unconstrained gradient algorithm (B.4) as long as $\theta \in \mathcal{C}^0$, i.e. θ is in the interior of the constraint set $\mathcal{C}$. If θ reaches $\delta\mathcal{C}$, the boundary of $\mathcal{C}$ *and* is trying to exit the set $\mathcal{C}$, then $\nabla J(\theta)$ in (B.4) is replaced by the projection of $\nabla J(\theta)$ onto the supporting hyperplane at the point of contact. This ensures that θ does not leave the set $\mathcal{C}$ as the iterative optimization is being performed.

A mathematical expression for the required projection can be derived quite easily [19]. For instance, suppose we wish to minimize the cost function $J(\theta)$, $\theta \in R^n$ subject to the constraint $\theta \in \mathcal{C}$ where $\mathcal{C}$ is a convex set in R^n. Furthermore, suppose that $\mathcal{C}$ can be described as $\mathcal{C} = \{\theta \in R^n | g(\theta) \leq 0\}$ where $g : R^n \mapsto R$. Then using the gradient projection method, we obtain the following algorithm:

$$\dot{\theta} = \Pr[-\gamma \nabla J(\theta)] \tag{B.5}$$

$$= \begin{cases} -\gamma \nabla J(\theta) & \text{if } \theta \in \mathcal{C}^0 \text{ or if } \theta \in \delta\mathcal{C} \text{ and } -(\gamma \nabla J)^T \nabla g \leq 0 \\ -\gamma \nabla J + (\nabla g \nabla g^T)/(\nabla g^T \nabla g)\gamma \nabla J & \text{otherwise} \end{cases} \tag{B.6}$$

A detailed derivation of the above algorithm can be found in [19].

REFERENCES

1. Bhattacharyya S. P., Chapellat H. and Keel L. H., *Robust Control: The Parametric Approach*, Prentice Hall, Upper Saddle River, NJ 1995.
2. Coppel W. A., *Stability and Asymptotic Behaviour of Differential Equations*, Heath and Company, Boston, Massachusetts, 1965.
3. Cutler C. R. and Ramaker B. L., "Dynamic Matrix Control — A Computer Control Algorithm," *AIChE National Mtg*, Houston, Texas, 1979; also Proc. Joint Aut. Control Conf., San Francisco, California, 1980.
4. Dahleh M. A. and Diaz-Bobillo I. J., *Control of Uncertain Systems: A Linear Programming Approach*, Prentice Hall, Englewood Cliffs, NJ, 1995.
5. Datta A., "Robustness of Discrete-time Adaptive Controllers: An Input-Output Approach," *IEEE Trans. on Automat. Contr.* Vol. AC-38, No. 12, 1852-1857, Dec. 1993.
6. Datta A. and Ochoa J., " Adaptive Internal Model Control: Design and Stability Analysis," *Automatica*, Vol. 32, No. 2, 261-266, Feb. 1996.
7. Datta A. and Ochoa J., "Adaptive Internal Model Control: H_2 Optimization for Stable Plants," *Automatica*, Vol. 34, No. 1, 75-82, Jan. 1998.
8. Datta A. and Xing L., "Adaptive Internal Model Control," *Adaptive Control Systems: Advanced Techniques for Practical Applications*, Eds. G. Feng and R. Lozano, 1998. (to appear)
9. Desoer C. A. and Vidyasagar M., *Feedback Systems: Input-Output Properties*, Academic Press, New York, 1975.
10. Doyle J. C., Glover K., Khargonekar P. P. and Francis B. A., "State Space Solutions to Standard H_2 and H_∞ Control Problems," *IEEE Trans. on Automat. Contr.*, Vol. AC-34, 831-847, Aug. 1989.
11. Egardt B., *Stability of Adaptive Controllers*, Lecture Notes in Control and Information Sciences, Vol. 20, Springer-Verlag, 1979.
12. Garcia C. E. and Morari M., "Internal Model Control — 1. A Unifying Review and Some New Results," *Ind. Engng Chem. Process Des. Dev.*, Vol. 21, 308-323, 1982.
13. Garcia C. E., Prett D. M. and Morari M., "Model Predictive Control: Theory and Practice — a Survey," *Automatica*, Vol. 25, No. 3, 335-348, 1989.
14. Goodwin G. C. and Sin K. S., *Adaptive Filtering, Prediction and Control*, Prentice Hall, Englewood Cliffs, New Jersey 1984.
15. Ioannou P. A. and Datta A., "Robust Adaptive Control: A Unified Approach," *Proc. of the IEEE*, Vol. 79, No. 12, 1736-1768, Dec. 1991.
16. Ioannou P. A. and Kokotovic P. V., *Adaptive Systems with Reduced Models*, Lecture Notes in Control and Information Sciences, Vol. 47, Springer-Verlag, New York, 1983.
17. Ioannou P. A. and Kokotovic P. V., "Instability Analysis and Improvement of Robustness of Adaptive Control," *Automatica*, Vol. 20, No. 5, 583-594, 1984.

18. Ioannou P. A. and Sun J., "Theory and Design of Robust Direct and Indirect Adaptive Control Schemes," *Int. Journal of Control*, Vol. 47, No. 3, 775-813, 1988.
19. Ioannou P. A. and Sun J., *Robust Adaptive Control*, Prentice Hall, Englewood Cliffs, New Jersey, 1996.
20. Ioannou P. A. and Tsakalis K. S., "A Robust Direct Adaptive Controller," *IEEE Trans. on Automat. Contr.*, Vol. AC-31, 1033-1043, Nov. 1986.
21. Kailath T., *Linear Systems*, Prentice Hall, Englewood Cliffs, New Jersey, 1980.
22. Kamen E. W. and Heck B., *Fundamentals of Signals and Systems Using Matlab*, Prentice Hall, Upper Saddle River, New Jersey, 1997.
23. Kharitonov V. L., "Asymptotic Stability of an Equilibrium Position of a Family of Systems of Linear Differential Equations," *Differentsial'nye Uravneniya*, Vol. 14, 2086-2088, 1978.
24. Kwakernaak H. and Sivan R., *Linear Optimal Control Systems*, Wiley Interscience, New York, 1972.
25. LaSalle J. P. and Lefschetz S., *Stability by Lyapunov's Direct Method with Application*, Academic Press, New York, 1961.
26. LaSalle J. P., "Some Extensions of Lyapunov's Second Method," *IRE Transcations on Circuit Theory*, 520-527, Dec. 1960.
27. Luenberger D. G., *Optimization by Vector Space Methods*, John Wiley & Sons, 1968.
28. Michel A. and Miller R. K., *Ordinary Differential Equations*, Academic Press, New York, 1982.
29. Middleton R. H., Goodwin G. C., Hill D. J. and Mayne D. Q., "Design Issues in Adaptive Control," *IEEE Trans. on Automat. Contr.*, Vol. AC-33, 50-58, 1988.
30. Morari M. and Zafiriou E., *Robust Process Control*. Prentice-Hall, Englewood Cliffs, NJ 1989.
31. Morse A. S., "Towards a Unified Theory of Parameter Adaptive Control - Part I: Tunability," *IEEE Trans. on Automat. Contr.*, Vol. AC-35, No. 9, 1002-1012, 1990.
32. Morse A. S., "Towards a Unified Theory of Parameter Adaptive Control - Part II: Certainty Equivalence and Implicit Tuning," *IEEE Trans. on Automat. Contr.*, Vol. AC-35, No. 9, 1002-1012, 1990.
33. Narendra K. S. and Annaswamy A. M., "Robust Adaptive Control in the Presence of Bounded Disturbances," *IEEE Trans. on Automat. Contr.*, Vol. AC-31, No. 4, 306-315, 1986.
34. Narendra K. S. and Parthasarathy K., "Identification and Control of Dynamical Systems Using Neural Networks," *IEEE Trans. on Neural Networks*, Vol. 1, 4-27, Mar. 1990.
35. Polycarpou M. and Ioannou P. A., "On the Existence and Uniqueness of Solutions in Adaptive Control Systems," *IEEE Trans. on Automat. Contr.*, Vol. AC-38, 1993.
36. Richalet J. A., Rault A., Testud J. L. and Papon J., "Model Predictive Heuristic Control: Applications to an Industrial Process," *Automatica*, Vol. 14, 413-428, 1978.
37. Rohrs C. E., Valavani L., Athans M. and Stein G., "Robustness of Continuous-time Adaptive Control Algorithms in the Presence of Unmodelled Dynamics," *IEEE Transactions on Automatic Control*, Vol. 30, No. 9, 881-889, 1985.
38. Rudin W., *Real and Complex Analysis*, McGraw-Hill, New York, 1974.
39. Soper R. A., Mellichamp D. A. and Seborg D. E., " An adaptive nonlinear control strategy for photolithography," *Proc. Amer. Control Conf.*, 1993.

40. Takamatsu T., Shioya S. and Okada Y., "Adaptive internal model control and its application to a batch polymerization reactor," *IFAC Symposium on Adaptive Control of Chemical Processes, Frankfurt am main*, 1985.
41. Tao G. and Ioannou P. A., "Robust Model Reference Adaptive Control for Multivariable Plants," *International Journal of Adaptive Control and Signal Processing*, Vol. 2, No. 3, 217-248, 1988.
42. Tsakalis K. S., "Robustness of Model Reference Adaptive Controllers: An Input-Output Approach," *IEEE Trans. on Automat. Contr.*, Vol. AC-37, 556-565, 1992.
43. Vidyasagar M., *Control System Synthesis: A Factorization Approach*, MIT Press, 1985.
44. Vidyasagar M., *Nonlinear Systems Analysis*, 2nd Edition, Prentice Hall, Englewood Cliffs, New Jersey, 1993.
45. Ydstie B. E., "Auto Tuning of the Time Horizon," *Proceedings of the 2nd IFAC Workshop on Adaptive Control*, 215-219, Lund, Sweden, July 1986.
46. Youla D. C., Jabr H. A. and Bongiorno J. J., " Modern Wiener-Hopf Design of Optimal Controllers - Part II: The Multivariable Case," *IEEE Trans. Automat. Contr.*, Vol. AC-21, 319-338, 1976.
47. Zames G., "Feedback and Optimal Sensitivity: Model Reference Transformations, Multiplicative Seminorms and Approximate Inverses," *IEEE Trans. on Automat. Contr.* Vol. AC-26, 301-320, April 1981.
48. Zames G. and Francis B. A., "Feedback, Minmax Sensitivity and Optimal Robustness," *IEEE Trans. on Automat. Contr.*, Vol. AC-28, 585-600, May 1983.

INDEX